SpringerBriefs in Applied Science

PoliMI SpringerBriefs

For further volumes:
http://www.springer.com/series/11159
http://www.polimi.it

Franco Caron

Managing the Continuum: Certainty, Uncertainty, Unpredictability in Large Engineering Projects

POLITECNICO DI MILANO

Springer

Franco Caron
Management Economics and Industrial Engineering
Politecnico di Milano
Milan
Italy

ISSN 2282-2577   ISSN 2282-2585   (electronic)
ISBN 978-88-470-5243-7   ISBN 978-88-470-5244-4   (eBook)
DOI 10.1007/978-88-470-5244-4
Springer Milan Heidelberg New York Dordrecht London

Library of Congress Control Number: 2013934795

Printed on acid-free paper

Springer is part of Springer Science+Business Media (www.springer.com)

Preface

Project Risk Management seems to be a mature discipline, based on standardized approaches developed by several institutions, such as the Project Management Institute. Nevertheless, a new challenge to the discipline derives from increasing complexity of Large Engineering Projects and their context. The publication of "The black swan" by N. Taleb represented a signal of the relevant and increasing role of unpredictability in planning and controlling projects. Project Risk Management deals only with anticipated uncertain events in order to identify response actions able to mitigate threats and enhance opportunities, this means that predictability is a requirement for risk management. The problem of preparing the project for the future including also the unforeseen emerges as a new challenge. Unpredictability can be considered as the extreme of the continuum "certainty uncertainty unpredictability" corresponding in Project Management terms to the continuum "project issues, project risks and unforeseen".

In this framework, the area covered by traditional project risk management, i.e. related to anticipated uncertain events, represents just a "grey area" interposed between a white area "certainty" and a black area "unpredictability". First, this book advocates that modern Project Management should deal with the whole continuum "project issues, project risks and unforeseen" and not just with a part of it. Secondly, since in the past different knowledge areas, such as traditional Project Management, Project Risk Management and Project Flexibility have been developed independently despite the strong dependencies existing between them, this book explores how to build an integrated approach addressing the whole continuum, which represents the true subject of the modern Project Management as an integrated discipline. Based on an extensive teaching, research and practical experience in the field of Large Engineering Projects, in particular in the oil and gas industry, a set of levers are identified in order to prepare the project for the future, such as traditional Project Management, forecasting capability, project risk management, real options, stakeholders management, responsive organizational model, etc. Along the project life cycle, the project team may use a mix of the above levers depending on the status of the project and its objectives. Since some levers may be unavailable or may be used to a limited extent depending on the constraints affecting the project, the choice of the mix of levers—what levers should be used and to what extent each should be used—will become the basic challenge for the project team in order to obtain an effective action on the whole continuum of "project issues, project risks and unforeseen".

The book may be used by students at graduate and master's levels, researchers exploring the recent trends in Project Management and practitioners interested in understanding how to prepare their projects for the future. Finally, I would like to thank Howard Evans for his constructive reviews of early drafts of this book.

April 2013   Franco Caron
                                                                  Politecnico di Milano

Contents

Chapter 1
Introduction

In Large Engineering Projects (LEPs), for instance those developed in the oil and gas industry, project effectiveness is a composite measure, combining economic performance, technical functionality, social acceptability, environmental sustainability, political legitimacy and economic development (Miller and Lessard 2000, 2001; Mc Leod et al. 2012). Such projects need large capital investment, have long time horizons and often use non-standard technology.

During the project life cycle, radical changes may occur, concerning one or more of the following: stakeholders, technology, investor needs, regulation, environmental concerns, market, personnel, top management commitment, economics, etc. On the other hand, LEPs are normally subject to extensive contractual constraints. As a matter of fact, the contract represents an implicit project baseline and, consequently, a source of constraints for the project execution plan. Moreover, LEPs are characterized by a large number of stakeholders involved in the decision making process (Flyvberg 2009). As a consequence, a LEP may be thought of as a lengthy decision making process, in which there are multiple times when decisions are required in order to respond to changing conditions.

In particular, LEPs are exposed to a high level of turbulence (Bosch-Rekvedelt et al. 2011; De Meyer et al. 2002; Loch et al. 2006), comprising different aspects:

- Uncertainty, i.e. lack of knowledge about the project and the future;
- Unpredictability, i.e. unexpected emerging situations;
- Ambiguity, i.e. possible different legitimate interpretations of the project situation at a given time.

In particular, unpredictability represents a significant challenge for LEPs due to increasing complexity in the projects and their environment (Makridakis et al. 2009; Wright and Goodwin 2009; Makridakis and Taleb 2009a, b; Taleb 2009, 2010).

Recent trends in the management of LEPs tend to extend the traditional view of the project, with a consequent increase of project turbulence, at least from the following three points of view:

- the project life cycle

F. Caron, *Managing the Continuum: Certainty, Uncertainty, Unpredictability in Large Engineering Projects*, PoliMI SpringerBriefs, DOI: 10.1007/978-88-470-5244-4_1, © The Author(s) 2013

- the external stakeholders
- the corporate level

Firstly, there is a trend toward a more extended view of the project life cycle, encompassing not only the execution phase, but also the previous phases, i.e. concept development and proposal management, and the subsequent phase, i.e. system operation and maintenance, that provides the benefits of the project for the owner's value chain. Secondly, not only the internal stakeholders should be taken into consideration, considering that internal stakeholders are normally linked to the project through contractual relationships, but also the external stakeholders, such as the authorities, media, local communities etc. In fact, both internal and external stakeholders may influence project success. Thirdly, the project should be managed considering its role in the project portfolio, which represents the link of the project to the corporate strategy.

As for project complexity (Williams 2005) and looking at the basic structure of LEPs, we can distinguish between an internal complexity which relates to the interdependencies between the different processes accomplished during the project life cycle and an external complexity which is related to the general environment of the project. As for the internal complexity in a LEP, we can distinguish three kinds of processes:

- operational processes
- managerial processes
- organizational processes

Each process represents a source of uncertainty and the interdependence between processes represents a source of complexity. Operational processes (i.e. design, procurement, construction, commissioning) determine the physical progress of the project, and generate the required deliverables, such as technical documents, purchase orders, deliveries at site, materials installed, testable systems. Managerial processes aim at planning and controlling the operational processes in order to focus the project on specific objectives such as cost, time and technical specifications. For instance, with reference to the PMI Body of Knowledge (Project Management Institute 2012), we can identify a sequence of managerial processes (i.e. initiating, planning, monitoring, controlling, closing up) for each knowledge area (cost, time, quality, risk, etc.). Organizational processes deal with human resources (selection, training, empowerment, coordination, compensation, etc.), since project objectives may be achieved only by the joint contribution of the people involved in the project. During the project development, operational, managerial and organizational processes are interwoven, increasing project complexity.

Focusing for instance on the operational aspects, LEPs are characterized by complex interdependencies between the various operational phases, i.e. design, procurement, construction, commissioning. The interdependencies among them are highlighted in Fig. 1.1, comprising three loops each describing a particular pattern of dependence.

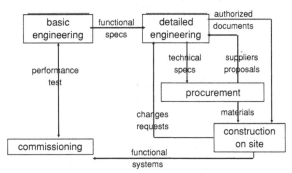

Fig. 1.1 Interdependencies between project operational phases

The inner loop is about the interaction between detailed engineering and procurement. Detailed engineering delivers the technical specifications necessary to obtain offers from suppliers and supports the procurement process until the issue of purchase orders. In turn, suppliers' offers are a source of information in order to improve the technical specifications. This kind of interaction implies a continuous exchange of information between detailed engineering and potential suppliers and, consequently, an overlapping between the two processes.

The middle loop is about the interaction between detailed engineering and construction on site. Detailed engineering provides the authorized technical documentation necessary to install the items and bulk materials provided by procurement and, in turn, field engineering provides the "as built" technical documentation, comprising the changes deriving from issues that emerged on site. This exchange of information allows the maintenance of a complete and updated description of the plant during the construction phase. Also in this case the interaction implies a suitable overlapping between detailed engineering and construction.

The outer loop is the most critical and represents the interaction between basic engineering and commissioning. Basic engineering defines the functional performance level expected from the system corresponding to the output of the project and hence commissioning should demonstrate the achievement of that performance level. The final test allows to verify whether actual performance levels meet contractual requirements. If not, radical and expensive changes may be required in order to avoid contractual penalties.

Throughout the interconnected loops, any unforeseen event may propagate across the overall project, determining an unpredictable impact. An increasing level of complexity tends normally to generate an increasing level of unpredictability.

In general, an overlapping between subsequent project phases is required in order to allow for the interaction between them. Figure 1.2 indicates the progress requested to the leading phase in order to allow for the start up of the following phase. If the overlapping becomes too strict, i.e. the progress of the predecessor

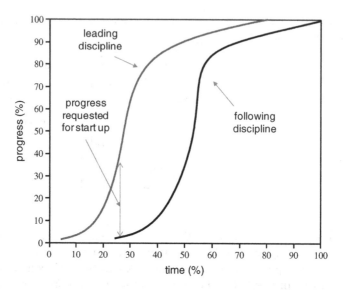

Fig. 1.2 Two overlapping project phases

is not significantly larger in comparison with the successor, we have a fast track project which is an extreme solution adopted in order to try to reduce overall project duration. This is a solution that may represent a risk for the achievement of project objectives since the successor process must develop its work without a sufficient contribution by the predecessor process. On the other hand, if the successor progress is significantly smaller than the predecessor, not only a delay of project completion is likely to happen but the exchange of information between the two processes may be difficult due to the lack of overlapping.

For instance, focusing on the engineering phase, the different disciplines involved, e.g. civil, mechanical, electrical engineering, should overlap, i.e. making a homogeneous progress, in order to allow for a continuous exchange of information, which is the prerequisite for a timely convergence toward an agreed upon and robust technical solution.

Also the interdependence between engineering and construction is very critical and requires an overlapping between the two phases. Firstly, at a given moment the progressive elaboration of a technical document by the engineering specialists should be interrupted, the document frozen, authorized for construction and sent to the site. If the document is frozen too early may be it is an incomplete document, otherwise, if it is frozen too late, a delay may derive for the construction process. In addition, the technical documents should be delivered to site in line with the construction sequence, i.e. engineering should be "construction driven", and hopefully the document should be sufficiently "robust" to avoid possible rework during construction.

Moving from the internal to the external complexity, political, economic, social, technological, legal, environmental interdependencies may exist between the project and the environment. For instance, the building of an infrastructure may suffer from the opposition of the local community.

In this complex context, LEPs try to stay on track, i.e. try to follow the original execution plan, while constantly adapt to emerging situations (Soderholm 2008; Geraldi 2008; Elms and Brown 2011), moving on the edge of chaos, between order and disorder, planned change and unplanned change (Geraldi 2010; Orton and Weick 1990).

Between contractual constraints and environmental turbulence, the project should find a trade-off between project stability, i.e. a high level of project robustness, and project adaptability, i.e. a high level of project flexibility (Soderholm 2008). Organizations adapt to environmental uncertainties through exploration and exploitation. Exploration includes the search for and experimentation with new approaches with the aim to find alternative solutions. Exploitation refers to efforts to gradually improve existing capabilities or processes aiming to capitalize on existing capabilities as much as possible (Liu and Leitner 2012). Both exploration and exploitation are likely to be needed by complex engineering projects to succeed.

Project robustness refers to the properties that enable the project to respond to the possible impact of uncertain events and so minimize the required changes on previous decisions, in particular on the project plan. Project flexibility refers to the properties that enable the project to reconfigure itself, introducing and exploiting degrees of freedom into the project plan and/or the project scope.

References

M. Bosch-Rekvedelt, Y. Jongkind, H. Mooi, H. Bakker, A. Verbraek, Grasping complexity in large engineering projects: the TOE (Technical, Organizational and Environmental) framework. Int. J. Project Manage. 29, 728–739 (2011)

A. De Meyer, C.H. Loch, M.T. Pich, Managing project uncertainty: from variation to chaos. MIT Sloan Manage. Rev. Winter 2002, 60–67 (2002)

D. Elms, C.B. Brown, Tales of the unexpected. Int. J. Risk Assess. Manage. 15(5/6), 387–399 (2011)

B. Flyvberg, Survival of the un-fittest: why the worst infrastructure gets built—and what we can do about it. Oxford Rev. Econ. Policy 25(3), 344–367 (2009)

J.G. Geraldi, The Titanic sunk, so what? Project manager response to unexpected events. Int. J. Project Manage. 26, 348–356 (2008)

J.G. Geraldi, L. Lee-Kelley, E. Kutsch, The balance between order and chaos in multi project firms: a conceptual model. Int. J. Project Manage. 28, 547–558 (2010)

L. Liu, D. Leitner, Sumultaneous pursuit of innovation and efficiency in complex engineering projects—A study of the antecedents and impacts of ambidexterity in project teams. Project Manage. J. 43(6), 97–110 (2012)

C.H. Loch, A. De Meyer, M.T. Pich, Managing the Unknown (Wiley, Hoboken, New Jersey, 2006)

S. Makridakis, N. Taleb, Decision making under low levels of predictability. Int. J. Forecast. 25, 716–733 (2009a)

S. Makridakis, N. Taleb, Living in a world of low levels of predictability. Int. J. Forecast. 25, 840–844 (2009b)

S. Makridakis, R.M. Hogarth, A. Gaba, Forecasting and uncertainty in the economic and business world. Int. J. Forecast. 25, 794–812 (2009)

Leod L. Mc, B. Doolin, S.G. MacDonell, A perspective-based understanding of project success. Project Manage. J. **43**(5), 68–86 (2012)

R. Miller, D. Lessard, *The strategic management of large engineering projects, shaping institutions, risks and governance* (MIT, Cambridge, MA, 2000)

R. Miller, D. Lessard, Understanding and managing risks in large engineering projects. Int. J. Project Manage. **19**, 437–443 (2001)

J.D. Orton, K.E. Weick, Loosely coupled systems: A re-conceptualization. Acad. Manag. Rev. **15**(2), 203–223 (1990)

Project Management Institute, *A Guide to the Project Management Body of Knowledge*, 5th edn. (PMI, Newtown Square, 2012)

A. Soderholm, Project management of unexpected events. Int. J. Project Manage. **26**, 80–86 (2008)

N.N. Taleb, Errors, robustness and the fourth quadrant. Int. J. Forecast. **25**, 744–759 (2009)

N.N. Taleb, *The Black Swan* (Random House, New York, 2010)

T. Williams, Assessing and moving on from the dominant project management discourse in the light of project overruns. IEEE Trans. Eng. Manage. **52**(4), 497–508 (2005)

G. Wright, P. Goodwin, Decision making and planning under low levels of predictability: enhancing the scenario method. Int. J. Forecast. **25**, 813–825 (2009)

Chapter 2
Large Engineering Projects Strategy

The project execution strategy of LEPs involves choices about issues such as partnership, outsourcing, modularity, etc. Projects must interact with their complex and uncertain environment and adapt their execution strategy to the ongoing changes (Artto et al. 2008). As a consequence, LEPs cannot be defined once and for all, rather they are shaped progressively from the initial concept by the dialectical interaction of stakeholders (Arrto et al. 2009; Miller and Lessard 2000, 2001; Koppenjan et al. 2011).

Turbulence in the project's environment involves uncertainty, unpredictability and ambiguity and implies a dynamic nature of the project's strategy, which can be better described as a process rather than as a front end prerequisite for project execution.

A change of project execution strategy may be a response to a changing internal/external context, such as the possible inability to maintain adequate financing, the trend of product price or widespread protests against the project (Nikander and Eloranta 2001; Pender 2001; Williams et al. 2009, 2010; Williams and Samset 2010).

In this context, uncertainty is mainly related to a lack of knowledge about the future development of the project, which requires the gathering of further information in order to better address emerging situations. From uncertainty may derive risks for the project (Perminova et al. 2008). Complexity is mainly related to the high number of elements involved in the project and the high number of interdependencies between them. Moreover, uncertainty may contribute to complexity, since for instance an insufficient definition of the scope of work may contribute significantly to the project complexity (Williams 1999). Complexity may represent the source of unpredicted events or conditions during the project life cycle. Ambiguity is mainly related to the coexistence of multiple interpretations of the project situation requiring a consensus building process based on the direct interaction of the stakeholders involved (Weick 1995).

As a consequence, at the heart of the choice of a project strategy we may find "certain" elements (issues and benefits deriving from the project's weaknesses and

F. Caron, *Managing the Continuum: Certainty, Uncertainty, Unpredictability in Large Engineering Projects*, PoliMI SpringerBriefs, DOI: 10.1007/978-88-470-5244-4_2,
© The Author(s) 2013

strengths respectively), "uncertain" elements (threats and opportunities deriving from uncertainty sources) and "unpredictable" elements, i.e. "black swans" (Taleb 2010) deriving from project's complexity. It should be noted that the traditional area covered by project risk management, i.e. related to anticipated risks, represents just a "grey area" interposed between the white "certainty" and the black area "unpredictability" (Floricel and Miller 2001). Since the three areas are normally addressed separately despite the strong dependencies existing between them, the possibility to develop an integrated approach should be explored.

In summary, the project strategy has to face a continuum of challenges ranging from issues/benefits (certainty) to unforeseen events (unpredictability) through anticipated threats/opportunities (uncertainty). The project strategy should aim not only at:

- using strengths to exploit opportunities
- using strengths to face threats
- controlling weaknesses to exploit opportunities
- controlling weaknesses to face threats

but also at preparing the project for the future i.e. for potential unexpected events and situations that may emerge during the project life cycle. In general, any strategic choice influence the overall continuum encompassing certainty, uncertainty and unpredictability. For instance, introducing some redundancy in the equipment in order to mitigate the risk of equipment failure causing a temporary work disruption also contributes to prepare the project to deal with unforeseen conditions by putting in place a reserve of resources.

At company level some strategies to cope with uncertainty may be identified (Engau and Hoffman 2011):

- Investigation, i.e. collecting additional information and drawing on professional expertise to be applied in decision making;
- Influence, i.e. manipulating circumstances or actors that constitute a source of uncertainty;
- Stabilization, i.e. implementing standard procedures or establishing long term contracts;
- Integration, i.e. restructuring business portfolio through divestitures, acquisitions and mergers;
- Flexibility, i.e. enlarging the range of strategic options to increase adaptability;
- Internal design, i.e. changing the organizational design by establishing modular structures, low degree of formalization, or decentralization;
- Postponement, i.e. deferring decisions and waiting for more certainty;
- "No regret" moves, i.e. executing activities that are advantageous regardless of how uncertainty resolves;
- Substitution, i.e. replacing uncertain decision criteria with assumptions derived from comprehensive consideration or detailed analysis;
- Simplification, i.e. reducing the number of uncertain factors considered in decision making;

- Cooperation, i.e. collaborating with suppliers, customers, or competitors, e.g. in research or production;
- Imitation, i.e. examining and copying the strategy of competitors;
- Withdrawal, i.e. exiting business in uncertain markets and focus on more predictable environments.

A set of strategic criteria suitable for addressing uncertainty in LEP's (Miller and Lessard 2001) may be summarized as:

- Identify/assess risks
- Transfer/hedge risks
- Diversify/pool risks
- Create real options
- Influence risk sources/mitigate risks
- Accept residual risks

The levers corresponding to the above criteria are knowledge management (identify/assess), contract and insurance (transfer/hedge), portfolio management (diversify/pool), project flexibility (real options), influence risk source (e.g. stakeholders), reduce risk exposure (mitigate) and provide a contingency reserve (accept residual risk).

A more comprehensive approach, aiming to cope with the overall continuum of "certainty-uncertainty-unpredictability", may be applied to LEPs, based on the following strategic levers:

- Apply Project Management processes
- Improve forecasting capability
- Enhance project robustness/flexibility
- Introduce real options
- Allocate, share, transfer risk
- Diversify, pool, escalate risk
- Mitigate major risk
- Accept residual risk (with contingency)
- Influence project's stakeholders
- Develop a responsive organizational model

The subsequent chapters will analyze each of the above leverage, with particular reference to LEPs in the oil and gas industry.

References

K. Artto, J. Kujala, P. Dietrich, M. Martinsuo, What is project strategy? Int. J. Project Manage. **26**, 4–12 (2008)

Artto K, Lehtonen M, Aaltonen K, Aaltonen P, Kujala J, Lindemann S, Murtonen M (2009) Two types of project strategy–empirical illustrations in project risk management. *IRNOP IX*, October 2009, Berlin

R. Miller, D. Lessard, *The strategic management of large engineering projects, shaping institutions, risks and governance* (MIT, NY, 2000)

R. Miller, D. Lessard, Understanding and managing risks in large engineering projects. Int. J. Project Manage. **19**, 437–443 (2001)

J. Koppenjan, W. Veeneman, H. Van der Voort, E. Ten Heuvelhof, M. Leijten, Competing management approaches in large engineering projects: the Dutch Randstad Rail project. Int. J. Project Manage. **29**, 740–750 (2011)

I.O. Nikander, E. Eloranta, Project management by early warnings. Int. J. Project Manage. **19**(4), 385–399 (2001)

S. Pender, Managing incomplete knowledge: why risk management is not sufficient. Int. J. Project Manage. **19**, 79–87 (2001)

T. Williams, K. Samset, K.J. Sunnevag (eds.), *Making Essential Choices with Scant Information* (Palgrave Macmillan, United Kingdom, 2009)

T. Williams, O.J. Klakegg, B. Andersen, D. Walker, O.M. Magnussen, L.E. Onsoyen, *Early Signs in Complex Projects* (PMI Research and Education Conference, Washington, 2010)

T. Williams, K. Samset, Issues in front-end decision making on projects. Project Manage. J. **41**(2), 38–49 (2010)

O. Perminova, M. Gustafsson, K. Wikstrom, Defining uncertainty in projects—a new perspective. Int. J. Project Manage. **26**, 73–79 (2008)

T.M. Williams, The need for new paradigms for complex projects. Int. J. Project Manage. **17**(5), 269–273 (1999)

K.E. Weick, The vulnerable system: an analysis of the Tenerife air disaster. J. Manage. **16**(3), 571–593 (1995)

N.N. Taleb, *The Black Swan* (Random House, New York, 2010)

S. Floricel, R. Miller, Strategizing for anticipated risks and turbulence in large scale engineering projects. Int. J. Project Manage. **19**, 445–455 (2001)

C. Engau, V.H. Hoffman, Strategizing in an unpredictable climate: exploring corporate strategies to cope with regulatory uncertainty. Long Range Plan. **44**, 42–63 (2011)

Chapter 3
Large Engineering Projects: The Oil and Gas Case

Oil and gas projects are often indivisible investments, irreversibly set in a location for a specific use. Moreover, such projects require large capital investment and have long time horizons, involving as a consequence a high level of uncertainty. These characteristics require that projects find a balance between stability, i.e. securing a continuous flow of revenues in the long term, and enough flexibility to adapt to emerging conditions (Merrow 2011).

The life cycle of a typical project in the oil and gas industry may be divided into three main stages: exploration, development and operation. Exploration refers to looking for geographical areas offering opportunities, identifying promising reservoirs, estimating their potential and acquiring exploitation rights. Development refers to locating wells, defining production and storage processes and putting in place on/off shore production units. Operation refers to oil and gas extraction, processing and export.

A company, representing the owner of the projects and operating in the oil and gas industry, normally covers the whole exploration, development and production cycle, from exploring oilfields to extracting, producing, refining and distributing refined oil to the final customer. In this context, the typical project entails the following phases (see Fig. 3.1).

Each single phase of the project aims at achieving a specific objective:

- *Evaluation*: carrying out the feasibility study of the project concerning a previously identified opportunity and the evaluation of how well it is aligned to the business strategy;
- *Concept selection*: developing alternative concepts in terms of technical and economical solutions and choosing the alternative which maximizes the project value;
- *Concept definition*: developing the design and planning of the selected concept;
- *Execution*: executing the project while aiming to meet the project baseline;
- *Commissioning, Start-up and Performance Test:* preparing and completing the final test representing the prerequisite for the start-up of the operation phase (i.e. first oil).

F. Caron, *Managing the Continuum: Certainty, Uncertainty, Unpredictability in Large Engineering Projects*, PoliMI SpringerBriefs, DOI: 10.1007/978-88-470-5244-4_3,
© The Author(s) 2013

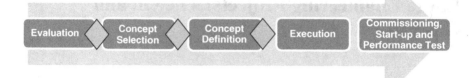

Fig. 3.1 Development project process

Figure 3.1 shows clearly that, when closing out a phase, there is a *gate* that represents a mandatory check point. Senior management reviews the project and gives the green light to proceeding to the next phase, otherwise the project is closed. The first gate is about the business opportunity and approves the business case. The second gate is about the product scope and approves the facilities plan and the third gate is about the execution project and approves the project cost and plan.

The focus on issues, risks and unforeseen events is extremely critical in the early stages of the project life cycle, when scant information is available and critical decisions must be taken. In the scheduling of an oil and gas project, there is a tendency to establish aggressive schedules, compared to other industries, in order to meet the *first oil* deadline (Merrow 2011) and improve the project cash flow. An aggressive schedule normally means that estimation of activity durations are excessively optimistic in comparison with schedule performance achieved in similar past projects. A *Schedule Aggressiveness Index* may be defined as the ratio between the schedule established at *financial investment decision* (FID, i.e. at the end of concept definition) and the schedule typically achieved for similar projects in the past (Merrow 2011). In fact, oil & gas projects are frequently characterized by low values of the index, often resulting in delays and an increase in investment during project execution.

The uncertainty of a project decreases with time as more knowledge is progressively revealed. For instance, at the project outset the reservoir may be affected by a high level of uncertainty in terms of size, properties, fluids composition and variability across the reservoir, temperature and pressure, presence of geological faults, etc. The characteristics of the reservoir and the extraction approach may not be known until surveying is completed. No drilling exploration is started until geological studies reveal a promising reservoir. No drilling exploitation is implemented until the reservoir appears to be commercially viable, and no large scale exploitation is implemented until the test wells confirm expected performance. Besides, other factors such as stakeholder's behavior, production profile, oil price, taxation level, etc. may change during the overall project duration, requesting a change of the original project plan.

As mentioned above, the overall decision making process is based on a sequence of decision gates, each allowing for the beginning of the subsequent project phase: evaluation, concept selection, concept definition, execution, operation.

As we move through the project life cycle from one gate to the next, the level of available knowledge increases and the corresponding level of uncertainty decreases. For instance, at the end of the concept selection stage, additional knowledge may become available about reservoir features such as oil viscosity, vertical heterogeneity, depth of the reservoir, etc.

Despite uncertainty, reservoir features and facility design should be considered in an integrated way in order to improve project performance. The evaluation of alternatives should take into consideration different criteria and their priority level. For instance, different plant configurations are possible depending on reservoir location and features: onshore plant, offshore platform, Floating Production Storage, Offloading ship connected by a flexible riser to the well, etc. Moreover, the choice between one or more units may depend on oil miscibility since the need to deal with different oils requires the use of more than one processing unit, independently from economies of scale considerations. The choice between building or leasing the plant may depend on the anticipated operational life of the reservoir on the estimated operational life of the units and the perspectives of re-use and conversion of the existing units. For instance, a long expected operational life makes the building alternative preferable.

In general, at each stage gate, the decision making process is characterized by:

- Information evolving as time goes on
- Multi criteria evaluation of alternatives
- Many stakeholders involved.

Given these characteristics, typical of complex projects, the decision making process may be aided by multi-criteria decision techniques, such as Analytic Hierarchy Process, allowing for a simultaneous consideration of the available alternatives, the different decisional criteria involved in any decision and the decision maker's preferences in terms of the weight given to each decision criteria (Guillaume et al. 2010). For instance, the decision making process may be "optimization" oriented or "robustness" oriented, assigning a higher level of preference, i.e. a higher weight, to one of the two approaches.

It should be noted that an optimization oriented decision making process implies the adoption of simplifying assumptions about the future context of the project (e.g. production profile, oil price, taxation level, etc.) in order to develop the project plan. Such assumptions may concern both external aspects, such as the oil market situation, or internal aspects, such as the well head temperature, which influences corrosion or wax formation. As a matter of fact, project assumptions may turn into project risks if they are subject to change, making the optimization oriented plan more exposed to risk.

If the decision making process is optimization oriented, the decision criteria are based on typical performance parameters such as cost, time, technical performance, etc. On the other hand, if the decision making process is robustness oriented, at least one new criteria should be added related to the capability of each alternative to absorb the possible impact deriving from the major risks hidden in the project assumptions.

For instance, focusing on the definition of the plant configuration, a critical decision node is represented by the choice of type and number of units (offshore platform, FPSO ship, etc.). If the decision making process is optimization oriented, considerations about economies of scale gain a dominant weight, supporting the choice of one unit, otherwise, based on service continuity requirements, the configuration based on two units becomes preferable.

References

M. Guillaume, G. Didier, L. Matthieu, Multi-criteria performance analysis for decision making in project management. Int. J. Project Manage. **29**, 1057–1069 (2010)

E.W. Merrow, *Oil industry megaprojects: Our recent track record* (Offshore Technology Conference, Houston, 2011), pp. 2–5

Chapter 4
Project Management

Traditional Project Management represents an organized way of dealing with the typical challenges stemming from uncertainty and complexity (Laufer et al. 1996).

First of all, the traditional role of planning and control should be considered. It is impossible to approach a journey without figuring out some sort of schedule and a project may be thought of as a journey. Planning should be a continuous process during the project life cycle entailing a sequence of re-planning actions in order to face new emerging conditions ("Planning is everything the Plan is nothing") (Dvir and Lechler 2004). In fact, planning is a "scenario building" exercise, i.e. a way of anticipating possible future issues. For instance, a resource loaded schedule may indicate possible work overloads requiring outsourcing measures. From this point of view, the concept of "estimate to complete" represents the core of the forecasting process required by the feed forward control mechanism typical of project control, since only the actions affecting the estimated work remaining can influence project performance (see Chap. 5).

Three typical project management techniques may represent examples of effective ways of addressing uncertainty and complexity: Work Breakdown Structure (WBS), Rolling Wave Planning and Concurrent Engineering.

The WBS allows for a definition of the project boundaries, a breakdown of the project scope into a set of more manageable work packages and so reducing project complexity. The WBS gives at least three fundamental contributions to project control. First, it represents a rigorous basis for the management of the project in terms of time, cost, etc. Second, a "deliverable oriented" WBS aims to focus the project on the customer's requirements. Third, the WBS allows for maintaining the coherence between a holistic view of the project (the higher levels of the WBS) and a detailed view (the lower levels of the WBS) respectively, i.e. between safeguarding project complexity and introducing some level of simplification (Giezen 2012).

While maintaining the project's overall master plan as a fixed reference, a "Rolling Wave" approach allows the detailed plan to be progressively elaborated and adjusted in response to the emerging conditions. Based on the Rolling Wave

F. Caron, *Managing the Continuum: Certainty, Uncertainty, Unpredictability in Large Engineering Projects*, PoliMI SpringerBriefs, DOI: 10.1007/978-88-470-5244-4_4, © The Author(s) 2013

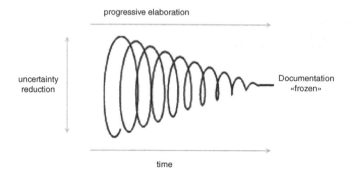

Fig. 4.1 Progressive elaboration

approach, project planning appears to be a threefold process: re-planning (taking into consideration emerging situations), detailing (extending the plan to the lower levels of the WBS) and freezing (authorizing the execution of operational tasks).

Differently from repetitive processes, the phases of a project are interdependent, since two overlapping project phases allow for exploiting the feedback from the following phase in order to improve the decision making process in the leading one (see Fig. 1.2). For instance, if engineering and construction phases overlap, possible issues which emerge on site should be taken into consideration during the elaboration of technical documents. Being bi-directional and continuous, communication will become a more fertile exchange of information that will assist in the earlier detection of faults, and will effectively reduce the future need for changes. In order to obtain this result, the progress of the subsequent phase should be in line with the progress of the preceding one, but, on the other hand, the engineering should follow a construction driven approach, meeting the requirements stemming from the construction sequence.

Focusing on the engineering phase, the development of the design process is based on a progressive elaboration of the technical documentation until the documentation is frozen (see Fig. 4.1). At each iteration, i.e. at each issue of a new version of the technical document, the comments coming from the different disciplines involved in the design process are included in the document, allowing for a progressively better definition of the deliverable and a corresponding reduction of the project's uncertainty.

As a consequence, a "Concurrent Engineering" approach needs the homogeneous progress of the different disciplines involved in the project in order to provide the consistency and completeness of the technical output and to avoid any possible rework or underperforming result. When the leading discipline achieves a sufficient progress all the involved disciplines may start. Through this approach, both "end" uncertainty (i.e. related to the deliverables) and "means" uncertainty (i.e. related to the processes) are solved gradually, and simultaneously allowing for taking agreed upon and robust decisions. Concurrent Engineering allows for minimizing the risk associated with the un-freezing of many hitherto frozen issues and solving any relevant interdependencies between the different disciplines involved.

In summary, Project Management, besides addressing through its basic processes the typical issues of the project such as time and cost, implies proactive measures able to cope with complexity and uncertainty affecting the project development. In particular, Project Management allows for a reduction of project's complexity (e.g. through the Work Breakdown Structure), the anticipation of future issues and the proactive intervention on the work remaining in order to meet the project objectives (through the project control process) and the progressive reduction of project's uncertainty through Concurrent Engineering.

References

D. Dvir, T. Lechler, Plans are nothing, changing plans is everything: The impact of changes on project success. Res. Policy **33**, 1–15 (2004)

M. Giezen, Keeping it simple? A case study into the advantages and disadvantages of reducing complexity in mega project planning. Int. J. Project Manage. **30**, 781–790 (2012)

A. Laufer, G.R. Denker, A.J. Shenhar, Simultaneous management: The key to excellence in capital projects. Int. J. Project Manage. **14**(4), 189–199 (1996)

Chapter 5
Improving the Forecasting Process in Project Control

Among the typical project management processes, planning plays a decisive role in reducing a project's uncertainty. Project planning may be thought of as resulting from the interaction of the project team with the project and the project context. Since uncertainty arises from a lack of knowledge, it is strictly linked to the inability of the project team to exploit all the available knowledge in order to anticipate the future development of the project (Williams and Samset 2010; Williams et al. 2009).

In fact, planning represents a forecasting exercise (Soderholm 2008). Developing a project plan is strictly related to making assumptions about the near and long term future, since the project plan reflects a summary of the assumptions taken about the future (Dvir and Lechler 2004). In a certain sense, the project plan can be thought of as representing a map of the future. Planning and forecasting are interrelated since they allow us, for instance, to fix milestones, e.g. critical dates for the project's stakeholders, in order to coordinate material/services deliveries, provided that forecasting lead times are consistent with delivery lead times. In particular, planning is necessary for each supplier in order to deliver their required contribution to the project in a timely way (Kleim and Ludin 1998).

Forecasting capability remains at the heart of project control. At a specific Time-Now (TN), a part of the work is completed (WC) and a part of the work is the Work Remaining (WR) that is still to be done. Based on the Earned Value Management System (EVMS) (Fleming 1992), the two components of the estimate at completion (EAC) are given by the Actual Cost (AC) of the WC and the Estimate To Complete (ETC) concerning the WR. Similar considerations may be applied to the estimate of Time at Completion (TAC). It should be noted that in the project control process the role of ETC is critical, since the only way to influence the overall project performance is to take actions affecting the WR. The information drawn from the ETC may highlight the possible need for corrective actions that may adjust the project plan (Anbari 2003; Christensen 1996). This approach corresponds to a *feed-forward* type control loop (see Fig. 5.1).

F. Caron, *Managing the Continuum: Certainty, Uncertainty, Unpredictability in Large Engineering Projects*, PoliMI SpringerBriefs, DOI: 10.1007/978-88-470-5244-4_5,
© The Author(s) 2013

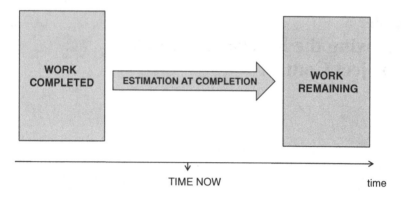

Fig. 5.1 Estimation at completion at time now (internal view)

As a consequence, during the project control process, the project manager plays a twofold role: the "historian", attempting to grasp the drivers that have determined the past evolution of the project, and the "wizard", attempting to foresee the future evolution of the project and to exploit all the lessons learned from the past (Makridakis and Taleb 2009; Makridakis et al. 2009).

This chapter will address the question of identifying the possible knowledge sources and how to integrate their different contributions in order to improve the forecasting capability during the project control process.

As the Project Management Institute (2012) has stated, the main processes involved in project management are initiating, planning, executing, monitoring, controlling and closing. In particular, Earned Value Management (EVM) represents an effective way of addressing the project control process. EVM is an efficient performance measurement and reporting technique for estimating the cost and time at completion (PMI 2012; Marshall et al. 2008). The following basic parameters are used in EVM, where TN indicates Time Now, i.e. the time along the project life cycle at which the project status is assessed:

- Planned Value (PV), the budget cost of work scheduled at TN;
- Earned Value (EV), the budget cost of work completed at TN;
- Actual Cost (AC), the actual cost of work completed at TN.

EVM was improved by Lipke (2002a, b, 2003), who introduced the concept of Earned Schedule ES for measuring schedule performance in time units and overcoming the flaws associated with a Schedule Performance Index SPI which is defined as the ratio between EV and PV, both expressed in monetary terms. Earned Schedule is the time at which the EV value, achieved at TN, should have been obtained according to the project baseline. The new Schedule Performance Index SPI(t) at TN, defined as the ratio between ES and TN, represents a more effective approach, since it avoids the problem of the convergence of the EV and PV values toward the BAC (Budget At Completion) as the project moves towards completion (Lipke 2006a, b).

The above three parameters and the ES value allow for the calculation of a set of indices and variances at TN. The most important of these are:

- Cost Performance Index

$$CPI = EV / AC$$

- Cost Variance

$$CV = EV - AC$$

- Schedule Performance Index

$$SPI(t) = ES / TN$$

- Schedule Variance

$$SV(t) = ES - TN$$

- Schedule Cost Index

$$SCI_{(t)} = CPI^* SPI(t).$$

Variances CV and SV(t) summarize the project's past performance during WC, whilst indexes CPI and SPI(t) may be used in order to highlight current trends and estimate the future performance during WR (Anbari 2003).

Many formulae to calculate the Estimate at Completion have been proposed during almost 50 years of EVM applications but none of them has proved to be consistently more accurate than any other one (Christensen 1993). In the standard approach, the estimation of final cost (i.e. EAC) and final duration (i.e. TAC, Time at Completion) are based on the following equations:

$$EAC = AC + (BAC - EV) / CPI_f \qquad (5.1)$$

where:

AC   Actual Cost at TN
BAC   Budget at Completion
EV   Earned Value at TN
CPI_f Cost Performance Index estimated for the work remaining (WR)

$$TAC = TN + (PAC - ES) / SPI_f \qquad (5.2)$$

where:

TN   Time Now
PAC   Planned at Completion, i.e. the planned duration of the project
ES   Earned Schedule at Time Now
SPI_f Schedule Performance Index estimated for the work remaining (WR)

It should be noted that future performance values may significantly differ from past performance (Davidson 1991). The new performance indices CPI_f and $SPI(t)_f$ have been introduced in Eqs. 5.1 and 5.2 with reference to WR and consider a possible evolution of the project which is different from that expected based on past performance. In summary, there are two different assumptions: either continuity between past and future performance or absence of continuity between them. While the generic indices CPI and SPI(t) are related to WC, CPI_f and $SPI(t)_f$ will be related to WR. In fact, relying only on past performance while developing a forecast could be misleading, since considering only past values of CPI and SPI(t) is similar to driving a car whilst looking just in the rear view mirror, so making it impossible to dodge the obstacles that may lie on the route ahead.

Both Eqs. 5.1 and 5.2 indicate that the values assigned to the performance indexes CPI_f and SPI_f play a critical role in order to obtain an accurate estimate of the final cost and duration. As a consequence, forecasting capability can be improved by utilizing all the available knowledge about the performance indexes CPI_f and SPI_f (Liu and Zu 2007; Goodwin 2005).

In general, the knowledge available to the project team may be classified in two ways: explicit/tacit and internal/external. Explicit external knowledge corresponds to data records about projects completed in the past, including measures of forecasting capability in terms of the difference between estimated and actual overall cost and duration. Taking into consideration past experience should mitigate possible "optimistic" bias in estimating future project performance (Lovallo and Kahneman 2003). Explicit internal knowledge corresponds to data records concerning the work completed, allowing for an evaluation of project performance at Time Now. Tacit external knowledge concerns the identification of similarities between the current project and some past projects in order to allow for the transferability of past data to the current project. Tacit internal knowledge is about possible events/trends affecting the project's work remaining.

In general, the basic approaches available in order to improve the forecasting process during project control may be summarized as follows:

- Pattern analysis; exploiting the identification of typical patterns, e.g. described in terms of S-curves characterizing the progress of similar projects and indicating, for instance, the time/cost necessary to achieve a certain milestone;
- Simulation of the future development of the project; provided a mathematical/logical model of the project can be developed allowing for a "what if" analysis of possible scenarios.
- Trend analysis; based on the extrapolation of the project performance using the actual trend at Time Now of a performance index, e.g. productivity;

Focusing on the trend analysis approach implemented in the framework of EVMS (see Eqs. 5.1 and 5.2), the basic knowledge sources available to the project manager for improving the estimate at completion at Time Now may be summarized as:

- data records related to the current project's Work Completed (WC);
- experts' subjective estimate about expected project performance in Work Remaining (WR);

- data records about final performance of similar projects completed in the past.

According to the threefold classification of the knowledge sources, three different approaches to linear trend analysis may be identified:

- utilizing data records related to WC, by extrapolating the current performance trend toward the future;
- adjusting the trend stemming from data records related to WC through experts' judgment estimating the expected performance during WR;
- integrating the internal view of the project, i.e. data records related to WC and experts' judgment related to WR, with data records deriving from similar projects completed in the past.

The data records related to the values of the performance indexes during WC may be expressed as:

- a cumulative value at Time Now, i.e. CPI and SPI summarize the overall behavior of the project during WC;
- a sequence of values CPI_m and SPI_m related to each unit time e.g. each month; in this case the recent values of the indexes are more sensitive to emergent trend.

In the first approach, the trend of the recent values CPI_m and SPI_m may be analyzed in order to estimate by linear extrapolation the value of CPI_f related to WR (see Eqs. 5.1 and 5.2)

In the second approach, considering all the information generated inside the project, the data records gathered during WC correspond to the explicit knowledge, and may be integrated with the experts' judgments corresponding to the tacit knowledge, in order to estimate CPI_f and SPI_f related to WR (Palomo et al. 2006). While data records are related to the past, experts' knowledge may address the future, avoiding the misleading effect of looking just backwards.

The Bayes Theorem represents a rigorous and formal approach allowing an update of a prior distribution, which expresses the experts' preliminary opinion through the data records gathered in the field. For instance, the project team may assume a prior estimate of the final budget overrun, based on subjective expectations about the development of the current project, and this prior estimate may be updated based on the actual performance of the current project at Time Now (Palomo et al. 2006). In a Bayesian framework, the experts' preliminary opinions, related to non-repetitive processes such as projects, are an example of subjective probability. Subjective probability is defined as the degree of belief in the occurrence of an event, by a given person at a given time and with a given set of information. It should be noted that increasing the level of knowledge available may modify the value of subjective probability assigned to a future event and to CPI_f and $SPI(t)_f$ (D'Agostini 1999).

The contribution given by tacit knowledge i.e. by experts about the future development of the project, may concern:

- the impact from drivers which explain the project development during WC, and also presumably affecting WR, i.e. what kind of plausible drivers may have generated the actual development of the project until Time Now and how they will also influence the future (e.g. schedule aggressiveness, engineering completeness, owner involvement, turnover in project leadership, anomalous low bid from subcontractor, unsatisfied stakeholders, new technology, project team integration, project team staffing, front end engineering adequacy, etc.) (Merrow 2011);

- possible behaviors of the stakeholders involved in the project, e.g. opportunistic behavior. It should be noted that in this case the focus moves from risk events to risk sources, i.e. to the stakeholders;
- certain/uncertain events or conditions affecting project performance during WR which may originate both internally and externally to the project. Certain events may include planned corrective actions or contractual constraints, while uncertain events, i.e. risks, may arise both in terms of threats (i.e. adverse weather conditions) or opportunities (i.e. more efficient solutions deriving from suppliers collaboration);
- weak signals indicating emerging situations possibly affecting project performance (scope changes, long lead time items delivery delay, long lead time items changes, permits timeliness, engineering sequence aligned with construction, rate of rework in construction, missing data, etc.) (Merrow 2011; Williams et al. 2012).

As for the third approach, beside the use of internal knowledge, both explicit and tacit, external knowledge related to similar projects completed in the past may also be used. The use of data records related to similar projects completed in the past has been introduced both with reference to the project outset in order to improve the estimate of the project budget, in particular when a proposal has to be prepared, and with reference to the project control process at a generic Time Now, in order to implement effective corrective measures.

Even though project management systems have been extensively implemented in recent years, project failures in meeting planned objectives are common, in particular in large engineering and construction projects such as in the oil and gas industry (Merrow 2011). However, it remains an open question whether these failures are due to a lack of project efficiency during execution or to a lack of forecasting accuracy during the planning phase. In the former case, both positive and negative deviations from the baseline should be expected, depending on the evolution of the project. However, a systematic overrun in terms of cost and time may be explained as a weakness of the forecasting process since the project's outset (Hogarth and Makridakis 1981).

Kahneman's studies (Kahneman and Tversky 1979) show that a major source of planning failure, which influences the accuracy of final cost and duration forecasts, is linked to an exclusively *"internal"* view approach, i.e. based only on data deriving from inside the current project. Subsequently, the focus has moved to the psychological and political factors affecting the project planning process (Lovallo and Kahneman 2003), and, in particular, two main sources of planning failure have been identified (Flyvberg 2006, 2009).

Firstly, the *cognitive biases*. These entail two major aspects: over-optimism, i.e. the common attitude to assess future projects with greater optimism than justified from previous actual experience, and anchoring, i.e. to deal with complex decisions by selecting an initial reference point (the anchor stemming from past experience) and anchoring the estimate onto it.

Secondly, the *strategic and political pressures*. These may typically emerge during proposal preparation. Indeed, the approval of a project pre-supposes a competition involving different proposals, which often causes a voluntary underestimation of cost and duration by the project proposers in order to make their own proposal as attractive as possible.

In response to the above reasons for failures in forecasting, the need emerges to exploit all the available knowledge during the planning process in order to minimize any bias effect. In fact, as shown in Fig. 5.1, the traditional control process often focuses only on data related to the current project, corresponding to an exclusively *internal* view (Flyvberg 2006). An integration is needed between the *internal* and the *external* view, the latter is based on knowledge related to projects completed in the past (see Fig. 5.2) (Flyvberg 2006). The same approach may be applied both at the project outset and at a generic TN during the project life cycle.

In fact, it may be assumed that the current project can be viewed as belonging to a cluster of similar projects that were completed in the past. Note that the selection of the cluster of similar projects is basically subjective since it depends on the similarity criteria adopted (Savio and Nikoloupolos 2011; Green and Armstrong 2007). Some cases, in fact, may express strong ambiguity. For example, if a company has to estimate the costs of an investment in a new technology and in an unfamiliar technological domain, should it take into account the set of highly innovative projects developed in different technological domains or the set of barely innovative projects but belonging to the same technological domain? Neither the former nor the latter option may be the best solution but both should be considered (Kahneman and Tversky 1979).

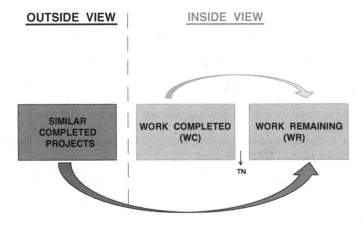

Fig. 5.2 External view and internal view

The process leading to the identification and use of a cluster of similar past projects comprises the following steps:

1. Recording the data related to past projects;
2. Classifying projects by similarity criteria;
3. Selecting the cluster of similar projects;
4. Analyzing the cluster in terms of relevant parameters, e.g. the final budget overrun.

Beside similarity criteria, the subjective assessment should also consider the trade-off between an extremely large number of projects, leading to the risk of including projects substantially different from the current one, and an extremely small number of projects, leading to a substantial loss of statistic significance. Also the transferability of past data to the current project and the possible presence of outliers in the distribution should be critically evaluated.

References

F. Anbari, Earned value project management method and extensions. Project Manage. J. **34**(4), 12–23 (2003)

D. Christensen, Determining an accurate estimate at completion. Nat. Contract. Manage. J. **25**, 17–25 (1993)

D. Christensen, Project Advocacy and the estimate at completion problem. J. Cost Anal. Manage. 35–60 (1996)

G. D'Agostini, Overcoming priors anxiety, in *Bayesian Methods in the Sciences*, ed. by J.M. Bernardo. Revista de la Real Academia de Ciencias, Madrid, 33(3) (1999)

P. Davidson, Is probability theory relevant for uncertainty? A post Keynesian perspective. J. Econ. Perspect. **5**(1), 129–143 (1991)

D. Dvir, T. Lechler, Plans are nothing, changing plans is everything: The impact of changes on project success. Res. Policy **33**, 1–15 (2004)

Q. Fleming, *Cost/Schedule Control Systems Criteria: The management guide to C/SCSC* (Probus Press Inc, Chicago, 1992)

B. Flyvberg, From nobel prize to project management: Getting risk right. Project Manage. J. **37**, 5–15 (2006)

B. Flyvberg, Survival of the un-fittest: Why the worst infrastructure gets built—and what we can do about it. Oxford Rev. Econ. Policy **25**(3), 344–367 (2009)

P. Goodwin, How to integrate management judgment with statistical forecasts. Foresight **1**(2005), 8–12 (2005)

K.C. Green, J.S. Armstrong, Structured analogies for forecasting. Int. J. Forecast. **23**, 365–367 (2007)

R.M. Hogarth, S. Makridakis, Forecasting and planning: An evaluation. Manage. Sci. **27**(2), 115–138 (1981)

D. Kahneman, A. Tversky, Intuitive prediction: Biases and corrective procedures. TIMS Studies in Manage. Sci. **12**, 313–327 (1979)

R. Kleim, I. Ludin I, *Project Management Practitioner's Handbook* (AMACOM 1998)

W. Lipke, *A study of the normality of Earned Value Management indicators* (The Measurable News, 2002a) Dec 2002, pp. 1–16

W. Lipke, Statistical process control of project performance. Crosstalk J. Def. Soft. Eng. **13**, 16–20 (2002b)

W. Lipke, *Schedule is different* (The Measurable News, 2003) Summer 2003, pp. 31–34

W. Lipke, in *Earned Schedule Leads to Improved Forecasting*. Proceedings of the 3rd international conference on project management (PROMAC 2006) Sept 2006, (2006a)

W. Lipke, *Statistical methods applied to EVM ...the next frontier* (The Measurable News, 2006b) Winter 2006, pp. 18–30

L. Liu, K. Zhu, Improving cost estimates of construction projects using phased cost factors. J. Constr. Eng. Manage. **133**, 1 (2007)

D. Lovallo, D. Kahneman, Delusion of success: How optimism undermines executives' decisions. Harvard Bus. Rev. **81**, 56–63 (2003)

S. Makridakis, N. Taleb, Decision making under low levels of predictability. Int. J. Forecast. **25**, 716–733 (2009)

S. Makridakis, R.M. Hogarth, A. Gaba, Forecasting and uncertainty in the economic and business world. Int. J. Forecast. **25**, 794–812 (2009)

R.A. Marshall, P. Ruiz, C.N. Bredillet, Earned value management insights using inferential statistics. Int. J. Proj. Manage. **1**(2), 288–294 (2008)

E.W. Merrow, *(Oil Industry Megaprojects: Our Recent Track Record* (Offshore Technology Conference, Houston, 2011), pp. 2–5

J. Palomo, F. Ruggeri, D. Rios Insua, E. Cagno, F. Caron, M. Mancini, On Bayesian forecasting of procurement delays: a case study. Appl. Stochastic Models Bus. Ind. **22**, 181–192 (2006)

Project Management Institute, *A Guide to the Project Management Body of Knowledge*, 5th edn. (PMI, Newtown Square, 2012)

N.D. Savio and K. Nikoloupolos, A strategic forecasting framework for governmental decision making and planning. *Int. J. Forecast.* Available on line (2011)

A. Soderholm, Project management of unexpected events. Int. J. Project Manage. **26**, 80–86 (2008)

T. Williams, K. Samset and K.J. Sunnevag (eds.) *Making Essential Choices with Scant Information*, Palgrave Macmillan (2009)

T. Williams, K. Samset, Issues in front-end decision making on projects. Project Manage. J. **41**(2), 38–49 (2010)

T. Williams, O.J. Klakegg, D.H.T. Walker, B. Andersen, O.M. Magnussen, Identifying and acting on early warning signs in complex projects. Project Manage. J. **43**(2), 37–53 (2012)

Chapter 6
Robustness and Flexibility

The traditional approach to Project Management focuses on the stability of the project plan as a critical success factor. However the increasing level of complexity and uncertainty in the business context requires a high level of adaptability to changes. Since emerging situations may jeopardize a company's assets, cash flow and reputation, a company that wishes to prepare for strategic surprises has two options. The first is to develop a capability for effective crisis management and the second approach is to treat the problems before they occur and thereby minimizing the probability of strategic surprises (Ansoff 1975). The former approach is typically "flexibility" oriented, the latter "robustness" oriented.

Project Robustness and Project Flexibility are typical responses to uncertainty, the former in a proactive way and the latter in a reactive way (Bernardes and Hanna 2008; Ross et al. 2008; Schulz and Fricke 1999). The former typically addresses anticipated risks; the latter develops the ability to react to unanticipated events or conditions affecting the project. Project Robustness aims to maintain the initial configuration of the project while facing changing conditions. Project Flexibility aims to modify the initial configuration of the project, e.g. the project plan, in order to adapt to the changing environment. In both cases, the set of measures taken to cope with anticipated risks and unanticipated events are at the core of project strategy (Artto et al. 2008, 2009; Floricel and Miller 2001; Morris and Jamieson 2005). Moreover, both Project Robustness and Project Flexibility allow for exploiting opportunities, both those identified during the early project stage, by putting in place specific risk response actions (e.g. a law change may be anticipated), or those that are unanticipated, by quickly adapting the project plan to the emerging opportunities (e.g. an unexpected more favorable composition of the site soil may represent a surprise during excavation activities allowing for a higher productivity) (Kolltveit et al. 2004).

As for stability, the main challenge for project planning is to make stable decisions that will stand the test of time. The main tool employed to achieve this target is to search for as much relevant information as possible before and during the decision making process (see Chap. 5) and to build into the plan some robustness

F. Caron, *Managing the Continuum: Certainty, Uncertainty, Unpredictability in Large Engineering Projects*, PoliMI SpringerBriefs, DOI: 10.1007/978-88-470-5244-4_6,

to face uncertainty. These steps allow the project to stay on a stable course and help to protect it against future uncertainties. In general, under conditions of uncertainty, managers do not attempt to make optimal decisions, they settle for robust i.e. stable decisions. "Robust" decisions should minimize changes to previous decisions due to emergent new conditions or, in any case, modify previous decisions at minimum cost. From this point of view, Project Risk Management (see Chap. 7) provides a fundamental contribution to improve project robustness, with the possible undesirable consequence that an over-commitment to preventive/protective strategies will produce an overconfidence on project success and deflect attention from the need to build flexibility into the project in order to cope with unanticipated situations (Loosemore 2006).

The central difficulty of a "robustness oriented" approach when defining a project strategy is the core premise that risks can be anticipated and the future may be forecasted, at least in probabilistic terms. This assumption may be realistic when sufficient information has been gathered to identify and assess uncertainty, even if this assumption is seldom true in the project early phase when the future remains mostly unpredictable.

In fact, it is inevitable that unanticipated events will occur in projects, consequently requiring a time pressured response. Typical categories of unexpected events are for instance:

- technical issues (e.g. performance test failure of a well known technical system),
- sponsor withdrawing support (e.g. project dismissed in senior management meeting),
- external events (e.g. force majeure events),
- resource change or constraint (key resource pulled off to work on other projects),
- human behavior (e.g. opportunistic behavior)
- project scope (e.g. major changes in scope) (Geraldi et al. 2010).

Contracts cannot protect projects from arbitrary decisions taken by government or from shifts in public opinion. Moreover, contracts between project's stakeholders can never address all possible emerging situations and rely on the underlying attitude of stakeholders to cooperate throughout circumstances that cannot be pre-specified.

Unpredictability which derives from external events and opportunistic behaviors of stakeholders calls for a high level of project flexibility and responsiveness, in order to allow for a quick reconfiguration of the project.

Focusing on flexibility, we can distinguish between product flexibility and project flexibility (Olsson 2006; Olsson and Magnussen 2007; Olsson 2008). Product flexibility tends to provide adaptability of the product to a changing demand whereas project flexibility tends to provide adaptability of the project plan to emerging situations. Both product and project flexibility may allow the project to absorb disruptive scope changes during project development. The need for project flexibility derives from the fact that important decisions affecting the project development are generally based on incomplete information since emerging events may radically change the assumptions which were originally adopted for project planning. The main requirements in the decision making process in order to achieve a high level of flexibility are:

- decisions should be postponed as long as the value of information remains high, maintaining future options for taking action when goals, preferences, alternatives and their consequences become clearer, in order to minimize the gap between the knowledge necessary to take the decision and the knowledge that is actually available (Davidson 1991);
- decisions should be taken in any case according to the lead time necessary to implement the corresponding actions.

For instance, freezing a technical document should be postponed as late as possible accordingly to the scheduling constraints deriving from the construction sequence. The decision making process implies a trade-off between gathering further information in order to get a more robust technical solution or to anticipate the completion of the deliverable to allow for the start up of the subsequent activity. Consequently, a suitable strategy may be based on postponing decisions as late as possible and taking robust decisions, i.e. changeable at low cost.

Three general strategies to exploit flexibility in the decision making process may be identified (Olsson 2006):

- Late locking
- Continuous step by step locking
- Contingency planning

The first strategy implies an iterative exploration of the project alternatives during the front end phase. Once the project is locked, as late as possible, the execution phase requires the stability of the project plan for success. The second strategy is based on a sequence of "decision gates", corresponding to the introduction of real options in the decision making process (see Chap. 2). The third strategy identifies a preliminary set of alternative plans that can be activated if needed.

Typical approaches that are aimed at improving project flexibility may be classified into: modularity, real options (see Chap. 8) and contingency planning (see conditional actions in Chap. 13). Modularity (Hellstrom and Wikstrom 2005) represents a powerful way of dealing with complexity by allowing for the breakdown of the project output into more or less independent sub-units and leading to a reduction of project's complexity.

With reference to LEPs, some typical measures may be investigated in order to cope with uncertainty and complexity, particularly improving both project robustness and flexibility (Ford and Bhargav 2006):

- Multiple subcontractors on site
- Redundancy of special equipment
- Mature technology
- No absolute beginners among suppliers/subcontractors

- Modular construction
- Preassembly at factory instead of assembly on site
- 4-D model (control of system interfaces, construction sequence,..)
- Field engineering
- Open purchase order associated with material take off of bulk material
- Long term supply frame agreement
- Hedge currency and interest rate exposure
- Etc.

With reference to flexibility, successful LEPs often adopt such approaches as (Miller and Lessard 2000):

- Functional specifications
- Participatory engineering
- Design-build contracts
- Robust design
- Standard technological solutions
- Media exploitation to prevent social opposition
- Contractual flexibility
- Contractual incentives
- Responsibility, risks, rewards allocation
- Secure access to knowledge and resources
- Alliances

The next two chapters will address project robustness and project flexibility respectively.

References

I. Ansoff, Managing strategic surprise by response to weak signals. Calif. Manage. Rev. **18**(2), 21–33 (1975)

K. Artto, J. Kujala, P. Dietrich, M. Martinsuo, What is project strategy. Int. J. Project Manage. **26**, 4–12 (2008)

K. Artto, M. Lehtonen, K. Aaltonen, P. Aaltonen, J. Kujala,S. Lindemann, M. Murtonen M., Two types of project strategy–empirical illustrations in project risk management, IRNOP IX, October 2009 (Berlin, 2009)

E.S. Bernardes, M.D. Hanna, A theoretical review of flexibility, agility and responsiveness in the operations management literature. Int. J. Oper. Prod. Manage. **29**(1), 30–53 (2008)

B.J. Kolltveit, J.T. Karlsen, K. Gronhaug. Exploiting opportunities in uncertainty during the early project phase. J. Manage. Eng. (2004). ASCE

P. Davidson, Is probability theory relevant for uncertainty? A post Keynesian perspective. J. Econ. Perspect. **5**(1), 129–143 (1991)

S. Floricel, R. Miller, Strategizing for anticipated risks and turbulence in large scale engineering projects. Int. J. Proj. Manage. **19**, 445–455 (2001)

D.N. Ford, S. Bhargav, Project management quality and the value of flexible strategies. Eng. Constr. Architectural Manage. **23**(3), 275–289 (2006)

J.G. Geraldi, L. Lee-Kelley, E. Kutsch, The balance between order and chaos in multi project firms: A conceptual model. Int. J. Proj. Manage. **28**, 547–558 (2010)

M. Hellstrom, K. Wikstrom, Project business concepts based on modularity—improved manouevrability through unstable structures. Int. J. Proj. Manage. **23**, 392–397 (2005)

M. Loosemore, Managing project risks, in *The management of complex projects: A relationship approach*, ed. by S. Pryke, H. Smyth (Blackwell Publishing, Oxford, 2006)

R. Miller, D. Lessard, *The strategic management of large engineering projects, shaping institutions, risks and governance* (MIT, Cambridge, 2000)

P.W.G. Morris, A. Jamieson, Moving from corporate strategy project strategy. Proj. Manage. J. **36**(4), 5–18 (2005)

N.O.E. Olsson, Management of flexibility in projects. Int. J. Proj. Manage. **24**, 66–74 (2006)

N.O.E. Olsson, External and internal flexibility—aligning projects with the business strategy and executing projects efficiently. Int. J. Proj. Organ. Manage. **1**(1), 47–64 (2008)

N.O.E. Olsson, O.M. Magnussen, Flexibility at different stages in the life cycle of projects: An empirical illustration of the freedom to maneuver. Proj. Manage. J. **38**(4), 25–32 (2007)

A.M. Ross, D.H. Rhodes, D.E. Hastings, Defining changeability: Reconciling flexibility, adaptability, scalability, modifiability, and robustness for maintaining system lifecycle value. Syst. Eng. **11**(3), 246–262 (2008)

A.P. Schulz, E. Fricke, Incorporating flexibility, agility, robustness and adaptability within the design of integrated systems—key to success? Digit. Avionics Syst. Conf. **1**, 1–17 (1999)

Chapter 7
Project Risk Analysis and Management

Focusing on a project's robustness, Project Risk Analysis and Management represents a typical proactive approach dealing with anticipated favorable/unfavorable events (see Chaps. 11, 12, 13, 14).

During the project's early phase, risks are firstly identified and allocated to the project stakeholders and secondly managed by each risk owner. In a LEP, risk allocation represents the first step in addressing uncertainty. A correct allocation of risks to project stakeholders is when each risk is borne by the stakeholder who is in the best position to manage the risk effectively and possibly tolerate its consequences. At project portfolio level, further risk minimization measures such as risk diversification, risk pooling and risk escalation can also be taken.

The contribution made by Project Risk Analysis and Management to a project's robustness is the exploitation of the knowledge available about the uncertainty aspects which could affect the future development of the project. The project's early phase should initially aim at reducing the uncertainty level, by:

- Clarifying goals, requirements, scope, etc.
- Making explicit the project assumptions
- Identify project stakeholders and their interests
- Exploring by simulation possible future scenarios
- Using available data records
- Eliciting experts' knowledge
- Improving the communication system (share project goals, project requirements, project scope, etc.)

In particular, attention should be devoted to the project assumptions. These are a normal way of simplifying future scenarios in order to proceed more easily with project planning. In turn, project assumptions, are key in identifying the underlying risks of the project which could arise from incorrect assumptions.

Response strategies to risks may be divided into three main areas:

- Developing specific response strategies for a few major risks;
- Providing a suitable contingency in order to absorb the joint impact of the residual risks on project performance;
- Cover intolerable risks by insurance.

In the first case, a response strategy to each major risk should be selected (avoid, transfer, mitigate for negative risk and exploit, share, enhance for positive risk) and implemented. For instance, common strategies are to invest in understanding risks, to share risks with partners and to shift risks to others through contract provisions where possible. Cost and time requirements deriving from the implementation of risk responses must be included in the project plan, so influencing the project budget and consequently the proposal to the client. In the second case the contingency, that is a part of the mark up added to the base cost, influences the price level offered to the client and, consequently, the competitive value of the proposal to the client. In the third case a tolerability threshold for the project in terms of the maximum acceptable loss should be determined pointing out the risks whose impact is not tolerable for the project and then must be covered by insurance. It should be noted that normally the perception of risk depends on the culture and situation of the stakeholders involved in the project (March and Shapira 1987; Loosemore 2006).

References

J.G. March, Z. Shapira, Managerial perspectives on risk and risk taking. Manage. Sci. **33**(11), 1404–1418 (1987)

M. Loosemore, Managing project risks, in *The management of complex projects: a relationship approach*, ed. by S. Pryke, H. Smyth, Blackwell Publishing, Oxford, (2006)

Chapter 8
Real Options

Contrary to the traditional system engineering approach, which optimizes a fixed design based on a set of fixed specifications, once the system and its objectives are defined, the Real Option approach recognizes that changes are inevitable over time and purposefully introduces flexibility into the project in order to address them. For instance, this can be achieved through a combination of a "wait and see" approach before uncertainty is resolved and a "partly reversible commitment" in order to minimize the cost of a possible withdrawal in case of project failure and possible reallocation of the committed resources to another project (Driouchi and Bennet 2012). In general, options such as wait, scale, switch, expand and abandon may be exploited for a Large Engineering Project (Zhao and Tseng 2003).

Real options are an established perspective on project flexibility with its roots in financial options theory. Project flexibility can be compared to owning an option for the right, but not the obligation, to take an action in the future. According to the real options model, uncertainty can increase the value of a project, e.g. the project cash flow, as long as flexibility is preserved and resources are not irreversibly committed. So keeping options open provides flexibility in the face of uncertainty allowing for an easier adaptation to emerging conditions (Brennan and Trigeorgis 2000; Dixit and Pindyck 2000; Miller and Waller 2003).

Focusing on project execution, among the different types of real options (option to defer, option to expand, option to abandon, option to switch, etc.), we can specifically focus on expandability which is the project's ability to add capacity. This is akin to a call option, and reversibility is the project's ability to undo its previous investment which is akin to a put option.

In financial terms, a call option, also known as "buy option", is a contract between two parties, the buyer and the seller. Here the buyer of the option has the right, but not the obligation, to buy an agreed quantity of a particular commodity or financial asset (the underlying asset) from the seller of the option at a specified time in the future (the expiration date) for a certain price (the strike price). The seller is obligated to sell the commodity or financial asset should the buyer so decide. The buyer pays a fee (called a premium) for this right. For instance, if the

F. Caron, *Managing the Continuum: Certainty, Uncertainty, Unpredictability in Large Engineering Projects*, PoliMI SpringerBriefs, DOI: 10.1007/978-88-470-5244-4_8,

development of a new product is considered as a call option then the present value of the cash flow that may derive from the product's sale is equivalent to the market price of the underlying asset, and the development cost corresponds to the strike price of the option.

A put option (usually just known as a "sell option") is a financial contract between two parties, the seller and the buyer of the option. The seller has the right to sell the underlying asset to the buyer of the option for a specified price (the strike price) during a specified period of time. If the option seller exercises his right, the buyer is obligated to buy the underlying asset from him at the agreed upon strike price, regardless of the current market price. In exchange for having this option, the seller pays the buyer a fee.

When the option is exercised, the owner obtains a return equal to the difference between the value of the underlying asset and the strike price. The value of the option derives from the fact that no owner would exercise the option unless the return is positive, determining asymmetrical returns, since they may indefinitely increase but they can't become negative. That's the reason why the greater the volatility of the value of the underlying asset the greater the value of the option.

In terms of an investment, different types of options are available: options that hold and phase investment, that change the amount of investment (growth, scaling and abandonment options) or alter the form of involvement (switching options) (Ford et al. 2002). In Large Engineering Projects, investments are normally irreversible. On the contrary, the expandability option may be normally exploited, since the opportunity to postpone at least partly a financial commitment in order to learn more about the uncertain future, increases the likelihood of project success (Dixit and Pindyck 2000). For instance, breaking down the project into a sequence of modules may offer a sequence of options for expansion, by splitting and delaying the commitment as a function of the actual market trend (Caron and Comandulli 2010). In Fig. 8.1 the project modularity allows for splitting the decision making process into a sequence of steps, providing a progressive expansion of capacity and consequently a progressive financial commitment.

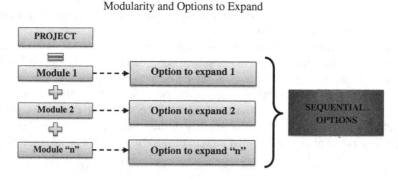

Fig. 8.1 Project modularity and sequential options

In general, real options theory offers a proactive approach for project managers, since many strategic decisions typically incorporate a series of steps—identify, analyze, rank, select, modify, cancel, etc.—in the light of available information. By introducing some flexibility into the project configuration, real options have the advantage of being able to postpone a decision, modify a decision in progress, reduce uncertainty surrounding it through acquisition of additional knowledge, and take advantage of volatility in its value. For instance, delaying procurement, such as postponing the issue of purchase orders for equipment, can add value to the project if future prices are uncertain or they happen to fall (Ford et al. 2002). In general an option provides an opportunity for the decision maker to take some action after uncertainties are revealed (Zhao and Tseng 2003).

As traditional cash flow analysis does not incorporate these degrees of freedom in the decision making process, it may not accurately estimate, e.g. in terms of Net Present Value, the relative value of an investment project offering real options. Real options must be considered explicitly in the simulation of the project cash flow, e.g. including the postponement of the expansion decision, and using the additional information gathered in the meanwhile, thus pricing the benefits of real options and improving the project's value (Caron and Comandulli 2010). At the heart of real options is the concept that being able to keep one's options open has value for the project, since it limits the damage in case of adverse unexpected events by avoiding any expansion choice or even withdrawing from the project.

References

M.J. Brennan, L. Trigeorgis (eds.), *Project flexibility, agency and competition: New developments in the Theory and Application of real Options* (Oxford University Press, Oxford, 2000)

F. Caron, M. Comandulli, in *Assessing Modularity as an Option for Expansion*. Proceedings of PMI global congress EMEA, Milan, 10–12 May 2010,(2010), pp. 1–7

A.K. Dixit, R.S. Pindyck, in *Expandability, reversibility and optimal choice in Project flexibility agency and competition*, ed. by M.J. Brennan, L. Trigeorgis (Oxford University Press, Oxford, 2000)

T. Driouchi, D.J. Bennet, Real options in management and organizational strategy: a review of decision making and performance implications. Int. J. Manage. Rev. **14**(1), 39–62 (2012)

D.N. Ford, D.M. Lander, J.J. Voyer, A real options approach to valuing strategic flexibility in uncertain construction projects. Construction Manage. Econ. **20**, 343–351 (2002)

K.D. Miller, H.G. Waller, Scenarios real options and integrated risk management. Long Range Plan. **36**, 93–107 (2003)

T. Zhao, C.L. Tseng, Valuing flexibility in infrastructure expansion. J. infrastruct. syst. **9**(3), 89–97 (2003)

Chapter 9
Stakeholders as Uncertainty Sources

LEPs are not defined once and for all, rather they are shaped progressively from an initial concept by the dialectical interaction of stakeholders (Arrto et al. 2008; Miller and Lessard 2001), and hence specific attention should be devoted by management to the attitude of the various stakeholders. It should also be noted that the dialectical interaction of the stakeholders may generate unpredictable consequences (Cooke-Davies et al. 2007) and changes of stakeholders' interests may result in the re-planning of the project (Soderholm 2008).

The management of the project is based on the inextricable interweaving of project control processes and stakeholders relational processes. At a higher level, project stakeholders may be defined as organizations or groups that have an interest or a functional role in the project and can contribute to, or be impacted by, the outcomes of the project (Project Management Institute 2012). From this point of view, projects may be described as a coalition of interest groups characterized by a political interaction (Newcombe 2003). Examples of project stakeholders can be sponsors, managers, suppliers, subcontractors, partners, clients, shareholders, financial institutions, insurance companies, governments, labour unions, mass media, pressure groups, consumers, local communities, etc.

Project management therefore, has a twofold focus: on the project objectives using the traditional planning and control techniques, and on the stakeholders, with whose commitment the project objectives can be realised (Leybourne 2010). Projects may fail because project management does not take the interests, objectives, expectations, attitudes, behaviours of stakeholders sufficiently into account.

In order to classify the project stakeholders, different criteria may be applied. Based on their level of involvement in the project, it is possible to differentiate stakeholders into either primary or secondary (Clarkson 1995). For instance, primary stakeholders should have a contractual or legal obligation to the project team (Cleland 1998), such as client, main contractor, suppliers, subcontractors, etc. Secondary stakeholders include, for instance, government (note that government can be a client as well), local authorities, media, consumers, competitors, local communities, etc. The current trend is toward an increasing role for the secondary stakeholders.

F. Caron, *Managing the Continuum: Certainty, Uncertainty, Unpredictability in Large Engineering Projects*, PoliMI SpringerBriefs, DOI: 10.1007/978-88-470-5244-4_9,

Projects can only be successful through contributions from stakeholders. And it is the stakeholders that evaluate whether the project is successful. In this context, the concept of project success appears to be inherently political. Various studies have illustrated the multi-dimensional nature of project success and described the different assessment criteria (Baccarini 1999; Shenar et al. 2001; Diallo and Thuillier 2005; Chan and Chan 2004; Bannerman 2008). For example, the following success criteria can be considered: the classical iron triangle (cost, quality, time), the product performance, the benefits to the organization of developing the project and the benefits for the local community (Atkinson 1999). Shenar et al. (2001) propose a different approach based on four dimensions: meeting time cost and quality requirements, benefits to the customer, benefits to the performing organization and preparing for the future. In either case, regardless of the success criteria adopted, it is important to emphasise how the definition of project success is highly dependent on the assessment expressed by the stakeholders involved (Stukenbruck 1986; Wideman 1998; Cicmil and Marshall 2005). In fact, each stakeholder—owner, manager, employee, supplier, etc.—expresses different expectations, and, therefore, has different criteria in order to assess project success. Moreover, these criteria may be implicit and changing over time. This is an enormous challenge for the project team. The underlying mechanism driving this political process, aimed at establishing the legitimacy of an interpretation of what is project success, is the stakeholders' pursuit of their interests leading to coalition formation or conflict (Kaplan 2008).

In order to describe the process linking each stakeholder to the possible impact on the project, firstly it should be noted that each stakeholder has general interests and, consequently, specific objectives for the project. Based on these objectives, the stakeholder formulates the corresponding requirements and creates expectations. Depending on whether or not these requirements are satisfied, the stakeholder shows different attitudes and behaviours, co-operative or obstructive with respect to the project. Note that a non-committed attitude might be sufficient to place the project in serious difficulty. Based on the available resources, the stakeholder can take actions so inducing significant consequences on project performance and success.

For instance, construction projects normally have some sort of impact on the surrounding environment, which could possibly create a conflicting relationship with local communities and environmental groups. The main interest of a pressure group, such as a pro-environment NGO, may be to be recognised by the authorities. If some aspect of the project concerns the group's social mission, i.e. the impact on the environment, or simply offers an opportunity to enhance its visibility, the group might explicitly propose an alternative technology, demand more stringent environmental controls, or request a meeting with managers in the presence of experts and authorities. As long as these requests remain unsatisfied, the group will threaten to mobilise all its resources, such as awareness campaigns, actions of opposition, demonstrations, blocks, and even alliances with the media, lawyers and researchers that increase its credibility. All these actions may lead to risks, which potentially have an impact on the project through delays or unexpected changes in the scope of work.

In this context, the salience of each stakeholder plays a significant role, since the greater the salience the greater the attention to be devoted to the stakeholder. The salience of the individual stakeholders can be assessed in terms of the presence of one or more of the following attributes: power, legitimacy and urgency (Mitchell et al. 1997). Power refers to the ability to influence the decision-making process; legitimacy refers to the legal context within which the project is developed, and urgency refers to the criticality and time sensitivity of the issues raised by the stakeholder. Furthermore, the level of salience usually depends not only on the individual characteristics of the single stakeholder, but more generally on the interactions with other stakeholders. In other words, power may derive from the position within the network of stakeholders, rather than from individual attributes (Rowley 1997; Neville and Menguc 2006).

Artto et al. (2008) argue that an understanding of the complexity of a project's stakeholder environment is an important feature of project strategy. Since LEPs are progressively shaped from an initial concept by the dialectical interaction of stakeholders, the assessment of the salience of the various stakeholders as well as the identification of suitable strategies to obtain their commitment to agreed objectives will play a fundamental role in determining project success and represents an essential part of the project management plan. Proactive management of the project stakeholders is required to reduce adverse behaviour that might adversely affect the project and to encourage active support of project objectives (Cleland 1998) (see Chap. 13).

Note that the stakeholders may be considered both as risk bearers and risk sources (Clarkson 1994, 1995; Post et al. 2002; Ward and Chapman 2008). For example, in international projects, some stakeholders can be identified as sources of political risk: host government, host society, inter-state relationships (Al Khattab et al. 2007). The host government may represent the source of a number of risks, such as taxation restrictions, currency inconvertibility, contract repudiation, import and/or export restrictions, ownership and/or personnel restrictions, expropriation and/or confiscation, industrial espionage, bureaucratic delays, etc. On the other hand, the host government may generate a series of opportunities, such as incentives and grants. In general suppliers, local community, partners, project team etc. are a source of risk, both in terms of threat and opportunity (Turner and Zolin 2012).

Based on the assessment of the potential threats/opportunities deriving from stakeholders on the project, either individually or as a group, different kinds of strategies may be used, such as understanding the stakeholder's drivers, communication, persuasion, negotiation, confrontation, etc. (Caron et al. 2010). The levers available to management to influence stakeholders can be of an organisational, contractual, institutional, political, economic or informational nature, such as participatory engineering, utilising Best Availability Technology (BAT) solutions, standardized solutions, media exploitation, contractual allocation of risk, introduction of incentives, development of a communication plan, creation of alliances, etc. For instance, lobbying may be a way for exercising influence for or against laws, regulations or trade restraints. In summary, influencing stakeholders' behaviour means, as a matter of fact, shaping the project itself and its environment (Jauch and Kraft 1986).

References

A. Al Khattab, J. Anchor, E. Davies, Managerial perceptions of political risk in international projects. Int. J. Project Manage. **25**, 734–743 (2007)

K. Artto, J. Kujala, P. Dietrich, M. Martinsuo, What is project strategy. Int. J. Project Manage. **26**,
4–12 (2008)

R. Atkinson, Project management: cost, time and quality, two best guesses and a phenomenon,
it's time to accept other success criteria. Int. J. Project Manage. **17**(6), 337–342 (1999)

D. Baccarini, The logical framework method for defining project success. Project Manage. J.
30(4), 25–32 (1999)

P. L. Bannerman, Defining success: a multilevel framework. in Proceedings PMI Research
Conference 2008, Warsaw, Project Management Institute, (2008)

F. Caron, F. Marini, F. Salvatori, Improving value with a risk based approach to stakeholder management. in *Proceedings of SAVE international 50th annual conference*, Long Beach, CA,
7–10 June, pp. 1–11, 2010

A.P.C. Chan, A.P.L. Chan, Key performance indicators for measuring construction success.
Benchmarking: Int. J. **11**(2), 203–221, Emerald Group Publishing Ltd

S. Cicmil, D. Marshall, Insights into collaboration at project level: complexity, social interaction
and procurement mechanisms. Build. Res. Inf. 33(6), 523–535

M.E. Clarkson, A risk model based of stakeholder theory. in Proceedings of the 2nd Toronto
Conference on Stakeholder Theory, Centre for Corporate Social Performance and Ethics,
University of Toronto, Toronto, (1994)

M.B.E. Clarkson, A stakeholder framework for analyzing and evaluating corporate social performance. Acad. Manag. Rev. **20**, 92–117 (1995)

D.I. Cleland, *Project Management Handbook* (Jossey-Bass, San Francisco, 1998)

T. Cooke-Davies, S. Cicmil, L. Crawford, K. Richardson, We're not in Kansas anymore, Toto:
mapping the strange landscape of complexity theory, and its relationship to project management. Project Manage. J. **38**(2), 50–61 (2007)

A. Diallo, D. Thuillier, The success of international development projects, trust and communication: an African perspective. Int. J. Project Manage. **23**, 237–252 (2005)

L.R. Jauch, K.L. Kraft, Strategic management of uncertainty. Acad. Manag. Rev. **11**(4), 777–790
(1986)

S. Kaplan, Framing contests: strategy making under uncertainty. Organ. Sci. **19**(5), 729–752
(2008)

R. Miller, D. Lessard, Understanding and managing risks in large engineering projects. Int. J.
Project Manage. **19**, 437–443 (2001)

R.K. Mitchell, B.R. Agle, D.J. Wood, Towards a theory of stakeholder identification & salience:
defining the principles of who and what really count. Acad. Manag. Rev. **22**(4), 853–886
(1997)

B.A. Neville, B. Menguc, Stakeholder multiplicity: toward an understanding of the interactions
between stakeholders. J. Bus. Ethics **66**, 377–391 (2006)

R. Newcombe, From client to project stakeholders: a stakeholder mapping approach. Constr.
Manage. Econ. **21**, 841–848 (2003)

S.A. Leybourne, Classifying improvisation: comments on managing chaotic evolution. in
Proceedings of 2010 PMI Research Conference, (Washington, USA, 2010)

J.E. Post, L.E. Preston, S. Sachs, *Redefining the Corporation: Stakeholder Management and
Organization Wealth,* (Stanford University Press, Stanford, 2002)

Project Management Institute, *A Guide to the Project Management Body of Knowledge*, 5th edn.
(PMI, Newtown Square, 2012)

T.J. Rowley, Moving beyond dyadic ties: a network theory of stakeholder influences. Acad.
Manag. Rev. **22**(4), 887–910 (1997)

A.J. Shenar, D. Dvir, O. Levy, A.C. Maltz, Project success: a multidimensional strategic concept.
Long Range Plan. **34**(6), 699–725 (2001)

A. Soderholm, Project management of unexpected events. Int. J. Project Manage. **26**, 80–86 (2008)

L. Stukenbruck, Who determines project success?. in *Proceedings PMI Annual Symposium*, Montreal 1986, (1986), pp. 85–93

R. Turner, R. Zolin, Forecasting success on large projects: developing reliable scales to predict multiple perspectives by multiple stakeholders over multiple time frames. Project Manage. J. **43**(5), 87–99 (2012)

S. Ward, C. Chapman, Stakeholders and uncertainty management in projects. Constr. Manage. Econ. **26**, 563–577 (2008)

R.M. Wideman, How to motivate stakeholders to work together, in *Field Guide to project management*, ed. by D.I. Cleland (Van Nostrand Reinhold, New York, 1998), pp. 212–216

Chapter 10
Project Organizational Model

As mentioned above, LEPs are exposed to a high level of turbulence comprising different aspects:

- Uncertainty
- Unpredictability
- Ambiguity

In particular, the degree of responsiveness of the project to unexpected events reflects the critical role played by the human factor in terms of adaptability to new game rules compared to the project outset, ability to interpret the emerging situation and to generate and implement a suitable response strategy (Saynisch 2010). It should be noted that in this case the project requires re-planning and learning instead of just triggering preplanned contingency responses as in Project Risk Management (Pitch et al. 2002). In this perspective, a project organizational model should focus on several points:

- Adaptive organizational model
- Leadership
- Mindfulness oriented organizational culture

In any organizational model, the members of the team should adopt an interdisciplinary approach so they can reconcile the objectives of their own discipline with those of the whole project. From an organizational point of view, projects are characterized by a high level of differentiation, both at higher and lower organizational level. At the higher level, several stakeholders, e.g. several companies, are involved in the project, each with diverse interests and cultural backgrounds. At the lower level, several organizational units contribute to the project progress, each representing diverse disciplines (legal, financial, technological, marketing, human resources, etc.). Nevertheless, the project requires a high level of integration between these different stakeholders, both at the higher and lower level.

In the project, the organizational model changes from a "mechanical" paradigm, based on rules and procedures, to an "organic" paradigm, based on direct

F. Caron, *Managing the Continuum: Certainty, Uncertainty, Unpredictability in Large Engineering Projects*, PoliMI SpringerBriefs, DOI: 10.1007/978-88-470-5244-4_10,
© The Author(s) 2013

interaction between the different roles involved in the project team. Organic structures are believed to be more effective than mechanical structures in uncertain environments largely because of their information processing capabilities, but at the cost of reducing the rules and procedures which are intended to make organization behavior more predictable. Moreover, the increasing level of interaction and communication in an organic structure allows for the generation of innovative ideas. The uniqueness of a project typically requires tailored solutions for unique problems. Since each organizational unit (or organizational role) involved in the project deals with a different aspect of the project's context, the overall adaptability of the entire project depends on the adaptability of each organizational unit to its particular aspect of the context. According to Ashby's law, a given level of diversity in the environment requires a corresponding level of diversity and independence across the different organizational units, since only diversity allows for an effective monitoring of the context in its different aspects. For instance, organizations based on loosely coupled systems allow for both differentiation and responsiveness across the different organizational units (Orton and Weick 1990). Each unit maintaining its culture but developing a dialectical relationship with the others and allowing for an effective organizational learning process. Shared values and mutual trust among team members are an absolute prerequisite for a high level of cohesion across the organization and help to improve the decision making process, particularly when dealing with unpredictable situations requiring a quick response (Godé-Sanchez 2010).

In reacting to uncertain and ambiguous situations allowing for different possible response strategies, the process of making sense of and interpreting the project's situation and building consent about a response strategy becomes critical (March 1978; Daft and Weick1984; Weick 1988, 1995a; Kaplan 2008; Alderman et al. 2005). Uncertainty and ambiguity can result in different interpretations about what is going on and what should be done. Not all the project team members may have homogeneous perceptions and similar evaluation criteria. In particular, weak signals must be interpreted, in order to take timely measures. For instance, a decrease in construction productivity may be interpreted as a radical shift in project performance or just a temporary downturn? To deal with uncertainty and ambiguity people interact, search for meaning, settle for plausibility and take action (Weick et al. 2005). The underlying mechanism driving this political process aimed at establishing a legitimate interpretation of the project situation is based on the interaction of the stakeholders, leading either to coalition formation or conflict (Kaplan 2008).

In this context, leadership plays a critical role. Two different views of the role of the project manager, as project leader, may be identified:

- In the first case, the project integration relies mainly on the Project Management System, as a set of detailed planning and control procedures concerning all the stakeholders/organizational roles involved in the project. In this case the project manager performs the role of "the supervisor" of the Project Management System;

- In the second case, a decentralized approach to project management may be implemented, based on relevant degrees of freedom to each organizational unit. In this case, there is no standard management system to be applied to all the organizational units, but for each unit a different organizational approach may be applied. The project manager, as project leader, undertakes the role of integrator of the various autonomous groups with different culture and focus, and becomes "the bridge" between diverse "languages", supervising the interface relationships between different organizational units.

The main advantages of the second case are:

– Safeguarding cultural diversity, as a way of allowing each organizational unit to monitor and adapt in a more effective way to its own environment, so improving overall project's responsiveness;
– developing innovation opportunities through the direct interaction of different organizational units, across the project.

As for "mindfulness", it should be part of the organizational culture as a pattern of shared beliefs and expectations that shape how individuals and groups act. Organizational culture may exert a centralized control over the dispersed activities by means of a handful of core values that are credibly established by top management, widely accepted by people in the organization, and used to interpret and express appropriate behaviors.

A "mindfulness" oriented culture is attuned to monitoring what is happening and is focused on grasping the weak signals that indicate a possible unexpected event. When facing complexity, "mindfulness" oriented organizations should focus on: preoccupation with failure, reluctance to simplify interpretations, sensitivity to operations, commitment to resilience and deference to expertise (Weick 1995b; Weick and Sutcliffe 2001, 2006). The typical organization's emphasis on procedures and preplanned contingency actions, embodies assumptions that weaken the ability to respond to the unexpected and foster new learning. The tendency to rely on procedures is part of a greater tendency to seek confirmation for existing expectations and avoid the change requests stemming from emerging situations. On the contrary, a "mindfulness" oriented culture focuses on anticipating, and becoming aware of, the unexpected and containing the unexpected when it does occur.

References

N. Alderman, C. Ivory, I. McLoughlin, R. Vaughan, Sense-making as a process within complex service-led projects. Int. J. Project Manage. **23**, 380–385 (2005)

R.L. Daft, K.E. Weick, Toward a model of organizations as interpretation systems. Acad. Manag. Rev. **9**(1), 284–295 (1984)

C. Godé-Sanchez, Leveraging coordination in project based activities: what can we learn from military teamwork? Project Manage. J. **41**(3), 69–78 (2010)

S. Kaplan, Framing contests: strategy making under uncertainty. Organ. Sci. **19**(5), 729–752 (2008)

J.G. March, Bounded rationality, ambiguity, and the engineering choice. Bell J. Econ. **9**(2), 587–608 (1978)

J.D. Orton, K.E. Weick, Loosely coupled systems: a re-conceptualization. Acad. Manag. Rev. **15**(2), 203–223 (1990)

M.T. Pitch, C.H. Loch, A. De Meyer, On uncertainty, ambiguity and complexity in project management. Manage. Sci. **48**(8), 1008–1023 (2002)

M. Saynisch, Beyond frontiers of traditional project management: an approach to evolutionary, self organizational principles and the complexity theory—results of the research program. Project Manage. J. **41**(2), 21–37 (2010)

K.E. Weick, Enacted sense-making in crisis situation. J. Manage. Stud. **25**(4), 305–317 (1988)

K.E. Weick, *Sense-making in organizations* (SAGE, Thousand Oaks, 1995a)

K.E. Weick, The vulnerable system: an analysis of the Tenerife Air Disaster. J. Manage. **16**(3), 571–593 (1995b)

K.E. Weick, K.M. Sutcliffe, *Managing the unexpected: assuring high performance in an age of complexity* (Jossey Bass, San Francicsco, 2001)

K.E. Weick, K.M. Sutcliffe, Mindfulness and the quality of organizational attention. Organ. Sci. **17**(4), 514–524 (2006)

K.E. Weick, K.M. Sutcliffe, D. Obstfeld, Organizing and the process of sense-making. Organ. Sci. **16**(4), 409–421 (2005)

Chapter 11
Introduction to Project Risk

All projects are risky ventures since they are unique and temporary undertakings based on assumptions about the future, affected by risks and subject to the influence of multiple stakeholders. The recent trend is toward an increase in project risk due to:

- the extended project life cycle, including the initial proposal phase and the final operation phase;
- the interdependence with other projects in the portfolio;
- the interaction with corporate strategy.

The overall project risk may also be influenced by some risk multipliers such as: size, complexity, innovative content, shortage of time, difficult locations, etc.

As explained in Chap. 1, a project deals with issues (certainty), risks (uncertainty) and unforeseen events (unpredictability). The risk differs from the issue since the risk is just a possible event whereas the issue corresponds to a certain problem currently affecting the project and often identified at the project outset. For example, an understaffed project represents an issue and a new supplier might represent a risk. A risk differs from an unforeseen event, i.e. a Black Swan (Taleb 2010) such as an unforeseen economical crisis in the euro zone, since the risk may be anticipated and proactively addressed whilst an unforeseen event may just allow for a reactive response. Examples of unexpected events in a large engineering project (LEP) may be geological issues, strong opposition from local communities, major delays in obtaining permits, turnover or relocation of key project team members, major change of client's requirements, change in regulations, client's default, economic crisis, etc.

The following chapters focus on the large grey area between the two extremes, issues and unforeseen events, the area covered by risk, i.e. possible favorable/unfavorable events affecting project performance (PMI 2012).

It should be noted that certainty, uncertainty and unpredictability may assume both a positive and negative aspect, e.g. in the case of risk, the threat can be viewed as negative (e.g. possible bad weather next week) and the opportunity as

F. Caron, *Managing the Continuum: Certainty, Uncertainty, Unpredictability in Large Engineering Projects*, PoliMI SpringerBriefs, DOI: 10.1007/978-88-470-5244-4_11, © The Author(s) 2013

positive (e.g. possible extension to other customers of the project results). In the same way issues and unforeseen events may determine either a positive or negative impact. Moreover, positive and negative impacts may be interweaved as a threat may turn into an opportunity since successfully dealing with an issue may increase company's reputation; conversely, an opportunity not exploited may turn into a threat since competitors may gain competitive advantage. This is part of the ambiguity that should be interpreted and dealt with by the project team, making sense of the project situation at a given time (see Chap. 10) (March 1978; Daft and Weick 1984; Weick 1988, 1995a; Kaplan 2008; Alderman et al. 2005).

The project is intrinsically risky since it represents a non-repetitive process where product and process are not completely defined at the project outset so generating a high level of internal uncertainty, for instance concerning the success of the final test. External uncertainty derives from the project context, e.g. from client's changing requirements.

Uncertainty can be viewed as the gap between the knowledge ideally required to successfully deal with a project and the knowledge actually available. It should be noted that the project outset is the most critical time, since relevant decisions related to project planning have to be taken whilst uncertainty is at maximum since no detailed analysis has yet been developed. In particular, during the life cycle of a large engineering project the bid/no bid decision is the most risky decision, since at that time the characteristics of the project may just be guessed given the relatively small amount of information available.

During the project life cycle the contract plays an important role since it divides the proposal management phase, when the project remains modifiable through negotiation between the parties, and the project management phase, when the project baseline is strictly defined, and consequently also project constraints are definitely fixed. In addition the amount of information available for project planning increases. In a sense the contract may be considered a draft of the project plan.

The typical subjects addressed in the contract are: technical specifications, price, payment terms, schedule, performance guarantees, warranties, limitation of liabilities, securities. Moreover, the contract represents the basic tool for allocating risks to the various project stakeholders, since each contract provision corresponds to the allocation of a specific risk to a project stakeholder. A typical example is given by the Lump Sum contract that allocates to the contractor the possible loss if there is a cost overrun, unless the loss derived from events occurred under the direct responsibility of the owner. The principle governing risk allocation is that each risk should be allocated to whoever is best able to manage it at the least cost. A correct risk allocation should reduce both the overall cost and the overall risk of the project. An improper risk allocation generates claims and may jeopardize the project success.

In summary, a distinction may be made between risk allocation and risk management. Firstly a "slice" of the overall project risk is allocated to each stakeholder, mainly using the contracts, and secondly each stakeholder has to analyze and manage his "slice" of project risk.

Project risk may be defined as "uncertainty that matters" (Hillson 2010), since the dynamics of risk envisages a possible event, i.e. the risk, determining an impact on project objectives, i.e. a significant deviation from expected performance. It is important to point out that "uncertainty that matters" may be analyzed, and modeled, in at least two different ways:

- risk event oriented,
- parameter variability oriented.

In the first case, the focus is on the *risk event* that is analyzed and managed individually; in this case each risk may be described in terms of occurrence probability and impact severity. In the second case the focus is on the *variability of a project parameter*, e.g. on the duration of an activity or the cost of an item, variability deriving from the joint impact of a set of micro-events influencing it. In this case, variability is normally described by means of a probability distribution. The second case is the typical "variability oriented" approach developed for instance by Program Evaluation Review Technique (PERT), assigning to each activity of the project network a duration distribution. In summary, the *risk event* based approach focuses on the single risk, whilst the *parameter variability* based approach may be applied to the estimation of the overall project risk.

It should be noted that parameter variability may be interpreted in two different ways: as the dispersion of data records (in case of repetitive processes allowing for collecting data records about activities durations) or as a lack of estimating confidence when planning the project (in case of non repetitive processes, such as the result of a football match, where the estimate corresponds to a bet on the future). In the former case a "frequentist" interpretation of probability is applied, i.e. stemming from frequency histograms based on data records, whilst in the latter case a subjective interpretation of probability is used, i.e. deriving from experts' judgments. In both cases, assuming that a standard distribution is used, e.g. a normal distribution, the corresponding standard deviation may be considered a measure of the uncertainty affecting the parameter considered, e.g. the activity duration.

The first approach, corresponding to an "event oriented" approach, will be developed in Chaps. 12 and 13 devoted to Project Risk Assessment and Management respectively, while the second approach "parameter variability oriented" will be developed in the Chap. 14 devoted to Project Risk Quantification. Note that in both cases, probability is used for expressing uncertainty.

In a Project Risk Analysis related to a LEP, the "event oriented" approach may be applied only to a limited number of risks, since the level of effort required for assessing and managing each risk may be significant, particularly if the response actions are expensive. As a consequence, the risks to be individually addressed during the Project Risk Analysis must be considered major risks, i.e. risks justifying the effort required to manage them. The "variability oriented" approach does not consider individually the extremely large number of micro events influencing a project, focusing the analysis on their joint impact affecting a project performance parameter, e.g. the project duration.

The above considerations allow us to introduce a first classification of project risks:

- Major risk, i.e. a risk to be addressed individually during the project risk analysis and management process. Normally, a major risk corresponds to an event that may jeopardize the project success if it occurs.
- Residual risk, the overall project risk not addressed in terms of individual risks but normally jointly covered by a contingency reserve.

Once the concept and a first classification of risk have been introduced, the next step is to identify the contribution given by Project Risk Management to Project Management. Both disciplines are characterized by a proactive approach. Project Management may be considered a way of dealing with uncertainty through the planning and control processes, anticipating the development of the project by means of project planning and taking suitable corrective actions by means of project control. On the other hand, the objectives of Project Risk Management are to take proactive measures in order to increase the probability and impact of positive events and decrease the probability and impact of events which will adversely impact the project objectives (PMI 2012). So what is the difference?

Each point estimate of a project parameter should be considered conditional on some simplifying assumptions that allow the elimination of the intrinsic project uncertainty. Project Risk Management aims to safeguard the project uncertainty moving from a point estimate of the project parameters, typical of traditional Project Management, to a distribution (or range) estimate, typical of Project Risk Management (see Fig. 11.1). Project Management processes based on point estimates normally assume an unrealistic degree of certainty about the project data. For instance, Project Scheduling assumes that the durations of the activities can be known in advance and therefore the completion date and the critical path can also be known. Project Risk Management describes the uncertainty of activity durations in terms of duration variability and/or possible risks affecting duration, consequently the overall project duration is also subject to variability. The more that

Fig. 11.1 From point estimate to distribution estimate

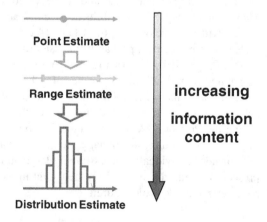

Point Estimate

Range Estimate

Distribution Estimate

increasing

information

content

uncertainty is recognized the more realistic will be the project plans and expectations of results.

Figure 11.1 indicates that moving from the point estimate of a generic project parameter through a range estimate towards a distribution estimate increases the knowledge available for project planning.

Moreover, project risk management (PRM) represents a formal process, based on a structured PRM System, entailing policies, roles, processes, techniques, tools, etc. PRM is part of project management, since the PRM plan should be integrated into the whole project plan, e.g. the planning of the response actions to risks may modify the cost and schedule baseline. On the other hand, the introduction of a contingency reserve may modify the total project budget. In both cases, risk gives a significant contribution to project cost and project price. In summary, PRM can be thought of as a bet, based on the expectation that the additional resources committed in managing the risks will be more than repaid in terms of a more effective control of project cost, schedule and technical performance.

Finally, as one key characteristic of Project Management is the progressive elaboration of the output deliverables, the completeness of the list of risks and the associated plan for responding to them also needs to be elaborated progressively. As long as there are uncertainties in the project, PRM has a continuing role. Finally, during project closure, risk related issues are addressed during the post project review, in order to contribute to organizational learning.

Before introducing the subsequent chapters dealing with risk identification, assessment and management, the concept of major risk, i.e. a risk addressed individually during the PRM process in order to identify and implement possible response actions, must be analyzed in greater detail. A project risk may be defined as an uncertain event or condition that, if it occurs, has a positive or a negative effect on at least one project objective (PMI 2012). The possible unfavorable effect is associated to a negative risk i.e. a threat, the possible favorable effect is associated to a positive risk i.e. an opportunity.

Typical risks in the engineering and contracting industry may be classified as: site risk, demand risk, force majeure risk, performance risk, latent defect risk, environmental risk, operational risk, change in law risk, contract risk, etc. Examples of major risks are: failure in forecasting market demand, opposition of local social groups, government decision to renegotiate the contract, enactment of a new law, bankruptcy of a subcontractor, technology failure, etc.

The risk, defined as an event, can be characterized by a set of elements:

- a source
- an occurrence time window
- an impact
- a project objective impacted

A risk source may be considered as an element of the project or the project context that may generate a threat/opportunity for the project. For instance, a supplier experiencing a work overload may create a delay in the delivery of a critical item. It should be noted that a risk source is in itself something neutral, as the above supplier should be considered, at least as a preliminary assumption, neither good nor bad, but during the project development, a chain of events may derive from the source leading to a positive or negative impact on the project. A risk source may be considered internal (technology, management systems, human resources, etc.) or external (site conditions, project stakeholders, weather conditions, etc.). The time window represents the interval during which the risk may occur, e.g. a construction risk during the construction phase. The impact on project objectives may generally be expressed in terms of cost, time or technical performance.

The risk may be described analytically based on these elements leading to a standard statement. For instance, because of using a new technology (source), unexpected system integration issues (event) may occur, during mechanical construction (time window), leading to rework (impact) affecting both cost and schedule baseline (project objectives).

Other examples of threats are:

- Because of rainy weather the excavations may be flooded during the foundation excavation work leading to work interruption so affecting completion date.
- Because of a lack of experience in the technology a misunderstanding of the customer's requirements may occur, during contract negotiation, leading to system underperformance and penalties.
- Because of conflicting tasks related to different projects a work overload for the electrical engineer may occur during the detailed engineering stage, leading to a delay of the final design review, affecting the start up of the construction phase.

Similarly, an example of opportunity:

- Because of production outsourcing, learning of new practices may occur during the procurement phase leading to increased productivity affecting financial performance.

References

N. Alderman, C. Ivory, I. McLoughlin, R. Vaughan, Sense-making as a process within complex service-led projects. Int. J. Project Manage. **23**, 380–385 (2005)

R.L. Daft, K.E. Weick, Toward a model of organizations as interpretation systems. Acad. Manag. Rev. **9**(1), 284–295 (1984)

D. Hillson, *Exploiting future uncertainty* (Gower, England, 2010)

S. Kaplan, Framing contests: strategy making under uncertainty. Organ. Sci. **19**(5), 729–752 (2008)

J.G. March, Bounded rationality, ambiguity, and the engineering choice. Bell J. Econ. **9**(2), 587–608 (1978)

Project Management Institute, A guide to the project management body of knowledge, 5th edn. PMI, Newtown Square (2012)

N.N. Taleb, *The black swan* (Random House, New York, 2010)

K.E. Weick, Enacted sense-making in crisis situation. J. Manage. Stud. **25**(4), 305–317 (1988)

K.E. Weick, *Sense-making in organizations* (SAGE, Thousand Oaks, 1995)

Chapter 12
Project Risk Analysis

Project Risk Analysis sets out to answer at the early stage of the project such questions as: what can go wrong? how can it happen? how likely is it to happen? what are the potential consequences if it does happen? In the succeeding phase risk management tries to answer such questions as: what can we do to respond to risks? what should we do? What should be the results of our actions?

The sequence of processes required to develop Project Risk Management (PRM) has been identified by Project Management Institute (PMI 2012) as:

- Risk management planning
- Risk identification
- Risk assessment
- Risk quantification
- Risk response planning
- Risk monitoring and control.

Firstly, Risk Management Planning implies that, in order to develop a Project Risk Management Plan, it is necessary that a PRM System exists, consisting of a set of policies, processes, procedures, roles, techniques, tools, templates, data records, lessons learned, etc. From this point of view, this book, in particular from Chaps. 11 to 14, may be read as describing the general structure of a PRM System and its components. The PRM System summarizes the overall knowledge accumulated by the company in developing past projects. This knowledge may be classified into the following categories: "tacit" (i.e. owned by the individual employees), "explicit" (i.e. stored in a data base) and "embedded" (i.e. included in policies and procedures). In a sense, the PRM Plan represents the application of the company PRM System to a specific project. The PRM Plan describes how the risk management processes should be carried out and how they should fit in with the other project management processes. The level and detail of processes, techniques and tools, the amount of resources allocated, the reporting requirements and the review frequency applied to PRM should be adapted to the characteristics of the specific project.

F. Caron, *Managing the Continuum: Certainty, Uncertainty, Unpredictability in Large Engineering Projects*, PoliMI SpringerBriefs, DOI: 10.1007/978-88-470-5244-4_12,
© The Author(s) 2013

For each PRM process, specific deliverables are generated: Risk Register (output of risk identification), Probability/Impact Matrix (output of risk prioritization), Risk Sheet (output of risk assessment) and Risk Response Plan (the output of the identification process of the response actions to be implemented). Understanding the content of each of these deliverables allows for a better understanding of the characteristics of the corresponding process. The typical contents of a PRM Plan may be summarized as:

- Introduction
- Project description
- Risk management methodology
- Risk management organization
- Techniques and tools
- Communication plan.

The second process in PRM is risk identification. When a risk is first identified potential responses should also be identified, otherwise the risk cannot be managed. The output of the process "Risk identification" is the Risk Register, listing all the significant risks considered to be relevant for the project. The choice of the number of risks to be analyzed is critical as there is a trade off between the completeness of the analysis and the effort required to perform it. Increasing the amount of risks to be assessed and managed individually may require an effort that is not compatible with the available resources. The choice between under-reporting, i.e. only considering a small amount of risks, and over-reporting, i.e. considering a large amount of risks, may also have psychological implications since at the outset of the project over-reporting may jeopardize the moral of the project team and their confidence in the success of the project. This correlation between the extent of the analysis and the corresponding level of effort, emphasizes the importance of the subsequent process of risk prioritization in order to limit the analysis to the major risks, i.e. the risks that individually considered may jeopardize the success of the project.

Before moving on to the techniques available for risk identification, the concept of risk as an event must be examined. In fact, the risk event must be described in detailed terms (who, what, where, how, when) in order to avoid any ambiguity about the occurrence of the event.

Before it occurs, an occurrence probability must be associated to the risk event. In general, this will be a subjective probability as it is related to a non repetitive event, which represents a typical situation in projects, that by definition, are non repetitive processes. This means that data records are not available or in any case not sufficient and consequently, probability represents just a degree of belief in the event occurrence. This accords with the definition of subjective probability, used for instance when betting on the results of a football match. At the end of the project, there are only two cases: the event occurred or it didn't occur. The same happens in a simulation process: an occurrence probability may be assigned initially to a given event but—in each simulation run i.e. in each actual "story" of the system under analysis—the event may happen or not.

First, when identifying project risks, a risk must be strictly defined as a relevant event, i.e. exercising an impact on project performance. Second, a major risk should not be a routine issue covered by other usual management systems, e.g. the quality control system requiring some planned performance tests. A lack of consistency in the technical documentation developed by the detailed engineering disciplines should be managed through a project quality management plan foreseeing some design review meeting, and not transferred to the risk management plan. In general, risk responses should not be just normal project management good practices, but each major risk should require a specific commitment aimed at reducing risk exposure. Third, a generic category of risks such as "delivery delays" cannot be considered an event, and consequently addressed as a risk, since this kind of risk, at least in part, occurs in every project, consequently it may be debatable later to what extent the risk happened. Fourth, variations of cost, time or technical performance should be avoided since they represent just the consequences of a risk event rather than a proper risk event. From this point of view a budget overrun doesn't represent a risk but is the impact deriving from a major risk or the joint effect of a set of minor risks. In the same way, site, security, contract etc. must be considered possible source of risks instead of being itself a risk. Besides, typical project constraints such as fixed time, fixed budget, fixed resources etc. cannot be identified as risks but correspond, as they are, to project constraints. In summary the risk, as an event, must not be confused neither with the risk source or the risk impact.

The techniques available in order to identify risks may be classified into three groups:

- using data records from similar projects (e.g. a check list);
- applying analysis techniques to the current project (Strength Weakness Opportunity Threat Analysis, Assumption Analysis, Stakeholder Analysis, Cause/Effect Diagram, Event Tree, etc.);
- Eliciting experts' judgment (Interviews, Brainstorming, Delphi Method).

Strength and Weakness Analysis corresponds to the "certain" aspects of the project, whereas Opportunity and Threat Analysis corresponds to the "uncertainty that matters" for the project. As for Assumption Analysis, an assumption corresponds to a scenario accepted as a valid base for project planning in order to reduce uncertainty. Due to changes that affect the project (stakeholders, technology, regulations, investor's needs, etc.) an assumption may subsequently be revealed as incorrect and turn into a risk for the project.

A common issue that often is overlooked or improperly addressed is how project assumptions and constraints are identified, reviewed and documented. Experience has shown that the process of how to manage project assumptions and constraints is essential to clearly understanding the project scope, minimizing project risk and fostering project success. Assumptions in project management refer to specific items that are considered to be true or certain when planning a project without necessarily having proof of it in reality. Unfortunately, even the most carefully considered assumptions typically carry with them a certain element of risk and if not properly addressed, result in a false sense of security in the project team.

Given the wide range of possible risks, the project risk manager should be first of all a knowledge manager since, both in the risk identification phase, and in the response actions identification phase, it is necessary to exploit all the knowledge available in the organization, both tacit and explicit. The project risk manager alone can't deal with all the different risks affecting the project but has to involve the experts in each knowledge area, playing the role of a "bridge" between the different specializations.

Project risks may be classified in different ways: major/residual, specific/systemic, insurable/not insurable, controllable/uncontrollable, tolerable/intolerable. A major risk is a risk which is addressed individually through specific response actions; otherwise residual risks are jointly covered by a contingency reserve. A specific risk is a risk affecting a single project, otherwise systemic risks (e.g. an economic crisis in the euro zone) may affect a set of projects. An insurable risk is a risk that may be covered by insurance (e.g. risk of damage of the plant during construction can be covered by All Risk insurance) otherwise (e.g. completion delay risk) the risk cannot be transferred to a third party and must be managed by the project team. A risk is controllable if a response action may be identified which is at least able to mitigate the risk, obtaining a reduction of the risk exposure greater than the cost of implementing the response action. A tolerable risk is a risk whose impact on the project performance may be acceptable to the project (or by the contractor developing the project) taking into account the maximum tolerable loss assumed by the project strategy. Specific and controllable risks may be addressed typically by a mitigation strategy. Specific and uncontrollable risks may be addressed by a contractual and insurance strategy. Systemic and controllable risk may be addressed by a portfolio diversification strategy.

A different approach to risk classification may be based on the risk source. Firstly, we may divide project risk sources into external and internal. External risk sources are related to the project context, entailing different aspects: political, economic, social, technological, legal, environmental. Examples of external risk sources are: government, government policy, trade barriers, exchange rates, inflation rates, materials or labour prices, transportation and communication infrastructures, etc. Possible risks deriving from risk sources such as the project stakeholders may be labour unrest or strikes, opportunistic behaviour by monopsonistic suppliers, default by client, etc.

As for the internal sources, we may consider three kinds of processes as sources of risk:

- operational processes (engineering, procurement, construction, commissioning, i.e. processes related to project progress),
- managerial processes (initiating, planning, monitoring, controlling and closing, i.e. processes related to project management),
- organizational processes, i.e. processes related to human resources.

For instance a change in laws or regulations represents an external political risk, while the possible turnover of qualified staff represents an internal organizational risk.

Such a classification may allow for a sufficiently complete view of the project and constitute the baseline for developing a Risk Breakdown Structure. A different approach to risk classification may be based on the project phase in which the risk materializes: feasibility risk, competitive bidding risk, engineering risk, procurement risk, construction risk, commissioning risk, operation risk. For instance:

- a feasibility risk may be not evaluating an alternative configuration of the system to be realized,
- a competitive bidding risk may be represented by a misunderstanding of the client's requirements,
- an engineering risk may correspond to a latent design defect,
- a procurement risk may involve the late delivery of a critical item,
- a construction risk may concern unforeseen geological problems,
- a commissioning risk may involve a test failure,
- an operation risk may concern a lack of availability of the system that is the output of the project.

A further approach to risk classification may be based on risk impact, introducing a distinction between internal impact (affecting the project baseline in terms of cost, time and technical performance) and external impact (affecting corporate reputation, competitive position, market share, etc.).

Once the project risks have been identified and classified, they may be reported in the Risk Register and then, assessed and prioritized. Eventually, suitable response actions may be identified and implemented for each major risk.

Typical risks that may be identified for a LEP are:

- Opposition from local community
- Technology novelty
- Obtaining Permits
- Environmental impact
- Lack of alignment between partners
- Lack of staff
- Lack of expertise
- Delivery delay of critical material
- Construction rework
- Stakeholder opportunistic behaviour
- Design change
- Laws and regulations change
- Industrial disputes
- Cultural differences

Risk assessment provides a better understanding of each individual risk and allows for the assignment of a priority level to each risk. Firstly, the basic features of each risk are highlighted: source, event, time window, impact on project objectives (see standard statement in the preceding chapter). Secondly, a further set of parameters may be addressed: probability of occurrence, severity of impact, controllability, tolerability, urgency, dependence relationships with other risks. In

Table 12.1 Risk sheet

• Project number
• Risk number
• Risk description
• Risk event
• Risk source
• Risk impact
• Risk time window
• Risk owner
• Risk classification
• Occurrence probability
• Impact severity
• Risk exposure
• Priority level
• Controllability
• Tolerability
• Urgency
• Dependence relationships
• Information sources
• Trigger events
• Response strategy
• Response actions conditional/unconditional
• Response actions implementation lead time
• Response actions expected effectiveness
• Secondary risks

particular the dependence between different risks must be addressed (Miller 1992). A set of correlated risks may determine something like a domino effect causing a relevant impact on the project performance, even if the single risks are not major risks (Miller 1992). All this information is summarized in the Risk Sheet devoted to each single risk (see Table 12.1).

The main parameters to be used to prioritize risks are: the probability of occurrence of the risk event and the impact severity, i.e. the risk impact on project performance in terms of cost, time, technical performance, cash flow, etc. A threat may impact a project in terms of additional cost, schedule delay or underperformance, whilst an opportunity may for instance determine a schedule enhancement, a cost saving and a performance improvement. The risk exposure may be determined as the product of these two parameters, probability of occurrence and impact severity. It should be noted that risk exposure represents the expected impact resulting from the risk. In general the expected impact may be described as the average impact obtained from the sum of the different impact values, each weighted by the corresponding probability value. In the simplest case we have just one impact value, e.g. 100 €, and the corresponding probability, e.g. 0.1, obtaining an expected impact, i.e. an exposure, equal to 10 €.

Risk exposure may be used as a criteria for prioritizing risks, since it represents a synthetic indicator of risk relevance. However, it would be a mistake to use risk exposure to evaluate the risk tolerance or to plan the contingency reserve. In fact, throughout the unique—i.e. non repetitive—history of the project, the risk will either happen or not happen, and in the former case the consequences will be given by the entire risk impact not by the exposure value.

Let us consider the following example where the risk corresponds to the possible test failure of a turbine, whose occurrence probability has been estimated to be 0.01 based on data records adjusted by expert's judgment. The impact severity may be estimated as the sum of 2 M€ due to rework and 10 M€ due to penalties for start up delay. So the risk exposure, i.e. the expected impact, corresponds to 0.12 M€. This value may represent a useful reference in order to prioritize this risk in comparison with other risks but it would be misleading to use this value as a reference for making decisions about the risk. In the real project if the risk occurs the actual loss will be 12 M€, so the decision to be taken is whether the company in charge of the project is prepared to tolerate such a loss, that may jeopardize not only the project success but also the company survival.

Instead of using a synthetic parameter such as risk exposure, risk prioritization may provide a distinction between the two dimensions, occurrence probability and impact severity, by means of the Probability Impact Matrix (see Fig. 12.1).

In defining the scales applicable to the two dimensions of the matrix, two different approaches may be applied for both parameters:

- a qualitative approach based on a sequence of very low, low, medium, high, very high levels, allowing just for a ranking of the risks;
- a quantitative approach for each risk based on a quantitative estimate of the probability of occurrence and the impact severity, allowing for a rating of the risks;

In both cases, looking at the Probability/Impact matrix (see Fig. 12.1), the major risks are those located in the right upper part of the matrix.

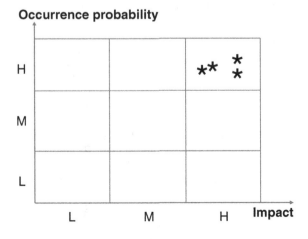

Fig. 12.1 Probability/impact matrix

In the quantitative approach, the probability scale may be defined by dividing the range 0–1 into several sections. It should be noted that in general the occurrence probability interval should be limited on the right side. For instance a probability value of 0.7 means that the risk event materializes in about two projects out of three, consequently corresponding more to a project issue than a project risk. A major risk should remain a rare event and a probability greater than 0.2 should be considered very high.

Defining a quantitative scale for the impact severity may be more difficult. Two questions should be addressed. First, how much impact would be completely intolerable in terms of a project objective (time, cost, technical performance, reputation, etc.) thus identifying the highest scale point. Second, how much impact would be completely acceptable (or in any case controllable through standard project control processes) thus identifying the lowest scale point. Assuming for instance that the project profit is 10 % of the contractual revenues, such a value may be taken as the highest scale point of impact severity, since an equivalent risk impact would determine the project failure.

Once the project's major risks have been detected and assessed in terms of occurrence probability and impact severity, the analysis of the project risk may be developed in a more detailed way. In fact, each risk event may be linked to a project element, e.g. a risk source (listed in the Risk Breakdown Structure) generating the risk, a Work Package (listed in the Work Breakdown Structure) affected by the risk impact and a Risk Owner (listed in the Organization Breakdown Structure) in charge of managing the risk. A Risk Owner is the member of the project team in charge of performing actions to implement a risk reduction strategy on a major risk. Other breakdown structures may refer to project deliverables, time windows, etc. In Fig. 12.2 each risk is defined by the intersection of a risk source (the lower level of the Risk Breakdown Structure) and a Work Package (the lower level of the Work Breakdown Structure), indicating where the risk originates and where it generates an impact respectively.

Fig. 12.2 Risk relationships

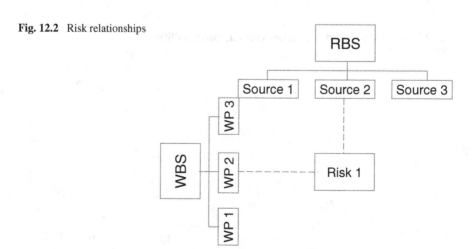

Based on this approach, it is possible to identify a set of risks having a project element in common, e.g. the set of risks assigned to a single organizational role, i.e. the Risk Owner, and summarize the corresponding risk exposure values, obtaining the overall Risk Load related to this project element, i.e. the overall risk exposure related to it. The well known rule "the expected value of the sum is equal to the sum of the expected values of the variables to be summed" can be applied. The concept of Risk Load allows for the identification of the critical elements of the project, e.g. a source generating a large number of risks, a Work Package affected by a large amount of impact or an organizational role in charge of a large number of risks.

It should be noted that if several risks arise from a common source, risk response actions may be more effective when focusing on the common source. Secondly, different risks stemming from the same source are normally correlated, since they tend to occur together. For instance an unforeseen increase in piping requirements initially underestimated may affect either procurement, transportation and construction work packages, determining a set of correlated impacts on project cost. Improving estimating accuracy of piping requirements, i.e. influencing the risk source, determines a mitigating effect on procurement, transportation and construction costs. On the other hand, the identification of joint effects on the same work package deriving from different risks, allows for the identification of a critical element of the project and the implementation of suitable protective actions.

References

K.D. Miller, A framework for integrated risk management in international business. J. Int. Bus. Stud. Second Q. **1992**, 311–331 (1992)

Project Management Institute, *A Guide to the Project Management Body of Knowledge*, 5th edn. (PMI, Newtown Square, 2012)

Chapter 13
Project Risk Management

The process of project risk management aims at implementing suitable response actions for each single major risk that has been identified (see Chap. 12). To do this, we need a systematic approach to provide a comprehensive set of effective actions otherwise we will only be able to rely on the experience and imaginative capacity of the risk owner.

The concept of a risk trigger allows for developing such a systematic approach. Between the risk source and the risk event there is a sequence of intermediate events, called risk triggers, which connect the two extremes, i.e. the source and the final impact. In a sense, we move from viewing the risk as an event to viewing the risk as a process, starting from the risk source and ending with the final event and the subsequent impact. The risk triggers play a twofold role. Firstly they indicate the approach to the final event, and secondly they represent a warning requiring attention and action from the management. In other words, the risk trigger represents the weak signal which anticipates major consequences. It is possible that at first the weak signals are vague, unclear and difficult to interpret (Nikander and Eloranta 1997; Weick 1995; Weick and Sutcliffe 2006; Williams et al. 2012), so contributing to the overall ambiguity of the project. For instance, a work overload at a supplier's factory represents the trigger of a possible delay in delivery. In a similar way, not answering a contractor's claim may evolve into a construction disruption.

Let us consider for instance a major risk such as an explosion of a confined cloud in a closed environment e.g. a household cellar. The risk source is the urban underground distribution network of methane gas. At the outset there is a change in the state of the source. A broken pipe may represent the first trigger i.e. the first step of the process moving toward the final event i.e. the explosion. The second trigger is the escape of the methane gas from the pipe, the third trigger the spreading of the gas underground in the directions allowed by the soil configuration. The fourth trigger is the entry of the gas into a cellar, the fifth trigger the formation of an explosive cloud, the sixth trigger an accidental ignition and, eventually, the explosion with the consequent impact.

F. Caron, *Managing the Continuum: Certainty, Uncertainty, Unpredictability in Large Engineering Projects*, PoliMI SpringerBriefs, DOI: 10.1007/978-88-470-5244-4_13, © The Author(s) 2013

Firstly, note that each trigger represents a warning i.e. a weak signal that may be captured so anticipating the successive steps of the risk process and implementing suitable actions aimed at interrupting the development of the risk process. Actions may be "unconditional" when implemented independently from the occurrence of the triggers, for instance:

- replace iron pipe with more stress resistant material (see risk source, i.e. methane distribution network),
- avoid any heavy traffic where iron pipe is laid at shallow depth (see trigger 1, i.e. broken pipe),
- use sufficiently large openings in the cellar.walls (see trigger 5, i.e. formation of an explosive cloud),
- use anti-explosion electrical devices (see trigger 6, i.e. accidental ignition).

On the other hand, actions may be "conditional" when they are activated by the occurrence of a trigger, e.g. an emergency team is called at first detection of gas. In general, a preplanned action (i.e. a contingency plan) may be "triggered" by the risk trigger. For instance, an evacuation plan allows for an organized escape from a building in case of fire. Note that the evacuation plan, that may be activated in case of emergency, represents a safety measure independent from the nature of the risk event (fire, explosion, loss of a toxic substance, etc.), i.e. it represents a multipurpose response to various events which have not been previously identified.

Secondly, note that after the occurrence of each trigger the probability of the final risk event increases. In Fig. 13.1 the above risk process is described by using an Event Tree model. Among the various models used in Safety Engineering, the Event Tree is based on a forward logic (from the triggers to the risk event and related impact), whilst a Fault Tree is based on a backward logic (from the risk event to

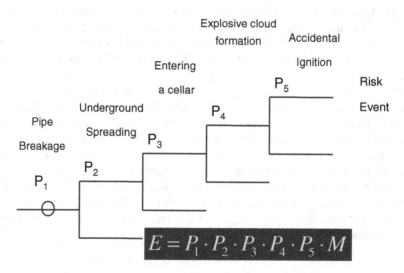

Fig. 13.1 Event tree describing the risk process

the possible triggers). The Event Tree indicates the sequence of the trigger events and each path along the tree leads to a possible final scenario. In the above example only one path leads to the risk event, i.e. to the explosion, the path connecting the sequence of the triggers that have been identified, but in general different risk scenarios may occur, each corresponding to a path of the tree. The risk exposure related to the risk event (e.g. the explosion in the above example) is given by the product of all the probabilities of occurrence of the sequence of triggers leading to the risk, times the impact severity (see Fig. 13.1). The formula indicates that, if one or more trigger events occur, since they become certain events and the corresponding probabilities become equal to one, the risk exposure consequently increases.

In the case of a construction project, an example of triggers may concern the risk of soil sinking under heavy loads due to the unknown subsoil characteristics. Anomalies during tunneling or excavation operations may represent the triggers that activate conditional actions such as a surface analysis or a deep seismic analysis in order to evaluate if reinforcement is required.

Another example may concern the risk process leading to a delivery delay at site and involving a typical sequence of possible triggers, such as partial delivery, delivery delay from the supplier, damage of the material, adverse weather during transportation by ship, customs problem, each trigger giving a contribution to the final delivery delay at site.

In general, the Event Tree represents an effective way of modeling the dynamics of a risk such as a delivery delay. To each trigger, an occurrence probability value may be assigned, allowing for the calculation of the probability of each final scenario by multiplying the probabilities of the triggers along the path associated to the scenario, depending on what kind of triggers occurred. Moreover, a value of the delivery delay to site may be calculated for each scenario. Note that the Event Tree is based on conditional probabilities, since the occurrence probability of each trigger may depend on the path considered, i.e. on what triggers occurred previously, consequently each path may have a different occurrence probability for the same trigger. This is an important feature of the Event Tree, allowing for modeling possible dependence relationships between the triggers included in the model. Eventually, for each scenario we obtain the corresponding probability and impact value. By summarizing the results, which are related to the different scenarios, the overall distribution of the probability of delivery delay may be obtained.

In summary, a systematic approach to identifying the risk response actions leads to the correct identification of the risk triggers together with a set of actions which aims to:

- avoid the risk process starting at the risk source,
- interrupt the risk process at each risk trigger.

This approach may be extended for instance to the analysis of the project stakeholders, since, with their unforeseeable behavior, they represent a significant source of risk for the project. The choice of a strategy aiming at influencing a stakeholder requires a systematic analysis of the dynamics of each major risk that it may generate. As explained above, risk can be described as a process that starts from a source (the stakeholder) and, through a chain of intermediate (trigger) events, creates an impact on project performance (Cagno et al. 2008). The main risk mitigation strategies envisage action on the source in order to modify its state or on the process, with the aim of breaking the chain of trigger events and so avoiding the final risk event. In order to describe the risk process linking the stakeholder to the final impact on the project, firstly it should be noted that each stakeholder has general interests and, consequently, specific objectives for the project. Based on these objectives, the stakeholder formulates requirements and develops expectations. Depending on whether or not these are satisfied, the stakeholder shows different attitudes and behaviors, co-operative or obstructive with respect to the project. Note that a non-committal attitude might be sufficient to place the project in serious difficulty. Based on the available resources, that he may control, the stakeholder can take actions so inducing significant consequences on project performance and success. Therefore, the risk process linking the stakeholder, as a risk source, to the final risk impact on the project may be described by the following sequence:

1. stakeholder's interests,
2. stakeholder's objectives,
3. stakeholder's requirements and expectations,
4. stakeholder's disposition and behavior,
5. stakeholder's available resources,
6. stakeholder's actions,
7. hence consequences for the project.

In fact, this sequence represents a dynamic progression from stakeholder's general interests to specific consequences for the project. The dynamic analysis of a generic risk that is sourced from a stakeholder can be used to identify the possible response actions (Cagno et al. 2008). At each step of the sequence, the project team can identify and take appropriate measures in order to break the chain of trigger events leading to the final risk event. In other words, stakeholder interests must not be allowed to evolve progressively towards actions that could put the project success at risk.

For instance, construction projects normally have some sort of impact on the surrounding environment and, consequently, a conflicting relationship with local communities and environmental groups. The main interest of a pressure group, such as a pro-environment NGO, may be to be recognized by the authorities. If some aspect of the project concerns the group's social mission, i.e. the impact on the environment, or simply offers an opportunity to enhance its visibility, the group might explicitly propose an alternative technology, demand more stringent environmental controls, or request a meeting with managers in the presence of experts and authorities. As long as these requests remain unsatisfied, the group will threaten to mobilize all its resources, such as awareness campaigns, actions of opposition, demonstrations, blockades, and even alliances with the media, lawyers and researchers that increase its credibility. All these actions may lead to risks, which potentially have an impact on the project through delays or unexpected changes in the scope of work. The more credible these actions are in terms of project impact, the greater the attention that management should pay to the NGO. Possible response actions implemented by the management could be an agreement with the NGO to monitor a particular pollutant, or the evaluation of different technologies during the development phase, so showing interest in environmental issues and avoiding obstructive behaviours. A further measure could be a meeting between management and the authorities immediately after any demonstrations or blockades.

Before addressing the strategies that may be applied in responding to risks, it is important to point out the different organizational levels at which project risks are dealt with:

1. the overall project involving different stakeholders, generally linked by contractual relationships;
2. the single stakeholder, e.g. the various contractors and subcontractors;
3. the project team.

At the first level, the overall risks affecting the project are allocated to the stakeholders and the contracts between the parties represent the means for sharing or transferring the project risks.

At the second level, the risks allocated to each single stakeholder, e.g. a company operating in the engineering and contracting industry, may be managed in the context of the project portfolio and the corporate strategy. Risk pooling and project diversification in terms of geographical area, technology, suppliers etc. allow the stakeholder to mitigate the impact both of single project risks and systemic risks affecting several projects such as a regional economic crisis. Moreover, a specific project risk may be escalated to the company level if it cannot be tolerated at project level but the project could play a strategic role, for instance by entering a new market.

At the third level, the project team ought to identify the most effective strategy to manage each risk. Based on the approach proposed by PMI Body Of Knowledge (2012), the available strategies may encompass, for threats:

- reduce uncertainty
- avoid threat
- transfer threat
- mitigate threat
- accept threat

and for opportunities:

- reduce uncertainty
- exploit opportunity
- share opportunity
- enhance opportunity.

For each selected strategy one or more actions may be identified and implemented. Each action should have a risk owner responsible for implementing and controlling it. Each of the above strategies will be discussed in the following.

Since risk may be defined as "uncertainty that matters" (Hillson 2010), uncertainty reduction represents a primary strategy aiming at enhancing the predictability of the future development of the project. For this purpose, all the available knowledge sources should be used since the effectiveness of each corrective measure is based on assumptions about the work remaining, which are in turn based on the best knowledge (tacit, explicit, embedded) available. Reducing uncertainty across the project may require different actions such as: clarify requirements and scope, verify assumptions, simulate possible scenarios, reuse past experience, acquire expertise, improve communication, etc. In general, collecting further information and improving the communication system contribute to the uncertainty reduction across the project. In the case of Large Engineering Projects (LEPs) a major objective is to understand the context of the project such as the country, the physical location, the political and regulatory institutions, the local community attitude, the local content, partners and external stakeholders, etc. Generally, a lack of knowledge exposes the project to foreseeable or unforeseeable shocks deriving from the context, for instance from the social context, such as a collective action blocking the construction process.

The second pair of risk strategies (avoid and exploit) have a strategic relevance for the project since they imply a change in the project configuration. For instance, the project scope of work may be reduced (eliminating a threat) or may be increased (capturing an opportunity). Typical strategic choices adopted in LEPs, such as partnership, outsourcing and modularization aim at avoiding threats and exploiting opportunities. For instance, modularization of the project output avoids the risk of a low productivity rate in the construction process at the site since the most of the work is completed in a controlled environment at the suppliers' factories obtaining a set of modules to be transferred to the site for the final assembly. Changing the vendor list, i.e. the project strategy, may also be a way of avoiding a threat and exploiting an opportunity.

The third pair of risk strategies (transfer and share) reflects the contractual strategy to be applied. A stakeholder may try to transfer a threat he is not able to manage or share an opportunity he is not able to exploit. The transfer of risk to other stakeholders may entail contractual delay penalties for the suppliers as long as it is clear that the stakeholder can effectively manage the risk. Cooperative strategies for sharing risk include long term agreements, alliances, joint ventures, technology licensing agreements and consortia. A particular type of transfer may involve an insurance company, transforming an uncertain cost i.e. the risk impact, into a certain cost i.e. the insurance premium. An All Risks insurance policy allows for the transfer of possible losses deriving from damage to the plant to an insurance company. In a similar way, hedging covers financial risks through derivatives (swaps, futures, options). In general, besides legal obligations concerning transportation of people and goods, a major intolerable risk constitutes a typical candidate for insurance.

The fourth pair of risk strategies (mitigate and enhance) plays an operational role by implementing prevention measures (reducing the probability of occurrence i.e. addressing the risk source) and protection measures (reducing the impact severity i.e. addressing the work element impacted) for the threats and vice versa for the opportunities. Modifying the project schedule in order to avoid excavation operations during forecasted bad weather represents a mitigation action oriented to reduce the probability of flooding open trenches. Alternatively, the availability of reserve resources, e.g. pumps, represents a mitigation action oriented to reduce the impact of the risk allowing for a quick return of the site to normal conditions. With reference to a construction project, a representative may be sent to the site if the productivity trend is decreasing (corresponding to a conditional action depending on the trigger decreasing productivity).

After addressing all the major risks, the remaining threats may be accepted without any specific response actions, but a contingency reserve must be added to the project budget to cover this residual risk. For instance, a completion delay risk may be accepted with a time buffer (i.e. with schedule contingency) or without a time buffer (i.e. without contingency, relying only on the total float available along the non critical paths).

Before proceeding with the implementation of a response action the controllability of the risk should be verified:

- the expected reduction of the risk exposure resulting from the implemented action should be greater than the implementation cost, i.e. the action should be cost effective;
- the action lead time, i.e. the time required for the action to be effective, should be consistent with the risk occurrence time window, i.e. the action should be time effective.

In general, the set of planned response actions should be selected taking into account effectiveness, timeliness and possible secondary risks.

Finally, all the response actions selected should be collected in the Response Action Plan that constitutes a part of the Project Execution Plan.

Unless the risk response plan is reviewed and updated regularly, the corresponding effort will not be adequately rewarded. The project risk management process requires a progressive review and update, since risk exposure changes constantly along the project life cycle. New risks may emerge and become "active", while other risks may disappear, either "deleted" by the response actions or "expired" since the occurrence time window was finished. Moreover, the effectiveness of the response actions may meet or not the expectations.

The occurrence of triggers has also to be monitored since they signal the approach to the final risk and may allow for the implementation of suitable "conditional" response actions. Moreover, secondary risks may materialize. Finally, the response actions should be monitored by considering if they were implemented and how effective they were. The project risk management process is an iterative process requiring a systematic review of the status of the project risks to be made at fixed intervals.

At project closeout, lessons learned should be reviewed by the project team which will identify the risks to be added to the Project Risk Check List, make changes in the Risk Breakdown Structure, consider response actions in planning of similar future projects, and make changes in the risk processes to improve their effectiveness. Moreover a proper view of residual risk will result in appropriate levels of contingency being set aside for similar projects in the future. In summary, the lessons learned from each project should allow for an improvement of the company's Project Risk Management System.

At project close out, it should be evaluated if any specific benefit to the project can be attributed to the PRM process in terms of robustness of the plan, reduced project duration, increased business benefits, client satisfaction, etc. In particular, the effectiveness of the project risk management process may be evaluated based on two elements:

- how comprehensive was the risk identification, pointing out the emergence and impact of unforeseen risks;
- how effective were the risk responses, pointing out the reduction of impacts resulting from foreseen risks.

In this framework, two KPI's may be based on the following ratios:

- risks identified materialized/overall risks materialized;
- overall actual risk impact/overall expected risk exposure.

The first KPI indicates the comprehensiveness of the risk identification process, whilst the second the effectiveness of the response actions identified and implemented. These KPI's, and others that may be added, allow for the improvement of the Project Risk Management System, pointing out possible critical areas requiring an intervention.

References

E. Cagno, F. Caron, M. Mancini, Dynamic analysis of project risk. Int. J. Risk Assess. Manag. **10**(1/2), 70–87 (2008)
D. Hillson, Exploiting future uncertainty, Gower (2010)
I.O. Nikander, E. Eloranta, Project management by early warnings. Int. J. Project Manage. **19**(4), 385–399 (2001)
Project Management Institute, *A Guide to the Project Management Body of Knowledge*, 5th edn. (PMI, Newtown Square, 2012)
K.E. Weick, The vulnerable system: An analysis of the Tenerife Air Disaster. J. Manag. **16**(3), 571–593 (1995)
K.E. Weick, K.M. Sutcliffe, Mindfulness and the quality of organizational attention. Organ. Sci. **17**(4), 514–524 (2006)
T. Williams, O.J. Klakegg, D.H.T. Walker, B. Andersen, O.M. Magnussen, Identifying and acting on early warning signs in complex projects. Proj. Manag. J. **43**(2), 37–53 (2012)

Chapter 14
Quantitative Analysis of Project Risks

Whilst the Risk Assessment process focuses on each single major risk and aims to identify suitable response actions, the Risk Quantification process provides insights into the joint impact of the overall uncertainty—both risk events and parameter variability—on the project. The focus moves from the single risk to the overall riskiness of the project. Quantitative analysis of the project risks has a two-fold objective:

- The evaluation of the overall project risk, considering the overall uncertainty stemming from both parameters variability and risk events, *before* implementing the response actions.
- The quantification of the contingency reserve, considering both the variability of the project parameters and the remaining impact stemming from major risks *after* implementing the response actions.

Exploiting the information about uncertainty, i.e. describing parameter variability by means of probability distributions, and using a quantitative technique such as Monte Carlo simulation as a tool for dealing with uncertainty, may provide more realism in the estimate of the overall project cost and duration than the traditional approach that assumes the activity durations and cost estimates are known with certainty. It should be emphasized that the probability distributions used in the quantitative analysis may be based both on data records (for repetitive processes) and the subjective degree of belief of the experts (for non-repetitive processes).

In order to proceed with the Risk Quantification process we need:

- a model of the project
- a set of input variables
- a quantitative technique
- a project objective

Focusing on the project schedule, the model adopted should be a network model which indicates the activities to be performed and the precedence

F. Caron, *Managing the Continuum: Certainty, Uncertainty, Unpredictability in Large Engineering Projects*, PoliMI SpringerBriefs, DOI: 10.1007/978-88-470-5244-4_14,

relationships between them. The input variables are the activity durations described by probabilistic distributions. As a quantitative technique Monte Carlo simulation has in recent years gained great success thanks to its ability to model any system, provided that it can be described in logical mathematical terms. The project objective may be represented by the completion date that corresponds to the output variable of the simulation model. The goal is to estimate the distribution of the output variable such as the project duration, as a function of the variability of the input variables such as the activity durations.

First of all we need to describe the project parameters in terms of probability distributions. When dealing with risk events the simplest distribution we may use is the binomial distribution which considers only two cases: (Y) the event happens or (N) it doesn't happen. When dealing with project parameters such as activity durations we may use standard distributions such as the Normal, the Beta (such as in PERT—Project Evaluation Review Technique) or the Triangular, etc. Sometimes these distributions have to be estimated by eliciting experts' judgment (this is the typical case of non-repetitive processes when data records are absent or insufficient). In these cases a well known approach is to ask the expert for "pessimistic", "most frequent" and "optimistic" values of the parameter and, based on these three values, to estimate the expected value and standard deviation (such as in PERT—Project Evaluation Review Technique) (Meredith and Mantel 2011).

The same approach may be applied to the calculation of the impact deriving from each risk event, since not only the risk occurrence but also the risk impact may be subject to variability. In general, a risk impact corresponds to a deviation from the expected performance, e.g. a budget overrun. Such deviation may derive either from an individual major risk (e.g. a final test failure) or from the joint effect of a set of correlated input variables (e.g. an increase of the pipeline requirements causing a set of correlated increases in procurement, transportation and construction costs, in turn determining a budget overrun) or from both of these causes.

A probabilistic distribution, for instance a distribution obtained from data records, may be reduced to a point estimate, e.g. for forecasting purposes, using different central values such as: the expected value, the modal value or the median value. The most important is the expected value, i.e. the weighted average of the values of the variable, weighted by their probability. The expected value allows for making direct calculations between the different variables since the expected value of the sum of the variables corresponds to the sum of the expected values of the individual variables (see for instance PERT). In project management other central values of the distributions are also used, even if not enjoying the same property of the expected value, i.e. the expected value of the sum corresponding to the sum of the expected values. The modal value, i.e. the most frequent value, is typically associated with the Critical Path Method, since the modal value seems to be the easiest value to be remembered when considering past similar projects. The median value, i.e. the value dividing the distribution into two equal parts, is associated with the Critical Chain approach since the median value represents the zero contingency value, i.e. the value with the same probability of overrun and under-run.

As mentioned above, moving from a point estimate to a distribution estimate of the project parameters provides a greater amount of information. By focusing on the overall distribution, and not just on its reduction to a point estimate, every distribution is described by at least two parameters corresponding to a central value, e.g. the expected value, and a dispersion value, e.g. the standard deviation.

It should be noted that the standard deviation, as a measure of variability, may assume a twofold meaning. In a "frequentist" view of probability, applicable to repetitive processes, such as a repetitive activity performed for similar projects e.g. trench excavation, the standard deviation describes the dispersion of the data records related to the variable, e.g. the dispersion of the past values of duration of the activity. In a "subjective" view of probability, applicable to non-repetitive processes where novelty dominates, the standard deviation describes the degree of belief of the project experts about the expected behavior of the variable e.g. the duration of the commissioning phase of a plant based on a completely new technology.

Until recently, the standard distributions used in modeling a project have been of the "thin tail" type, i.e. a very high impact has a negligible probability. For instance the normal distribution is almost completely concentrated in the range between minus/plus three standard deviations around the central value. In case of complex projects, possible non-linear effects should be taken into account, e.g. a delay in the delivery of a critical item may be progressively amplified by the complex interactions between the internal processes of the project causing eventually a dramatic completion delay. As the complexity rises, project elements and stakeholders interact in increasingly unforeseeable ways, multiplying the potential for a high risk impact. In this case "fat tailed" distributions allow the assignment of a realistic non-negligible probability to events involving a very high impact on the project performance. Considering the wider context of a project, a sudden 150 % increase of the oil price due to a Middle East crisis or a 50 % decrease of oil price due to a euro zone crisis are examples of low probability/high impact events that should not be neglected when running a simulation process.

Monte Carlo simulation is based on the "trick" of translating a probabilistic problem into a sequence of deterministic problems, i.e. of simulation runs. It should be noted that each simulation run represents a possible "story" of the project which is being analyzed, since at each run, each risk event happens or not, and for each project parameter a point estimate is sampled, i.e. casually extracted, from the corresponding distribution.

The sampling process of the distributions that describe the variability of the input variables is the mechanism that represents the link between the original problem that is probabilistic in nature and the single simulation run, that is deterministic in nature. At each simulation run, for each input variable a single observation is sampled from the corresponding distribution allowing for a deterministic estimate of the output variable (see Fig. 14.1). In a sense, simulation summarizes the impacts deriving from risk events and parameters variability, and estimates their joint effect on the variability of the project performance parameter, that may be assumed as an indicator of the overall project's riskiness, e.g. the variability of

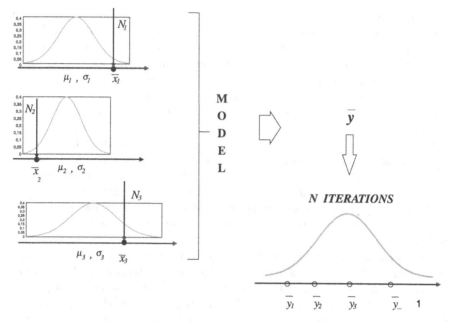

Fig. 14.1 Simulation process

the overall project cost. As a way of calculating the overall riskiness of a project, simulation corresponds exactly to the goal of the Risk Quantification process.

It should be noted that the result of a simulation process is a distribution of the output variable. The observations of the output variable obtained from the simulation runs may be summarized in a histogram from which a standard distribution may be obtained by making the best fit of the data. It would be a mistake to assume the central value of this distribution, e.g. the expected value, as an accurate forecast of the future behavior of the project. The simulation doesn't tell us which of the many possible stories will come true in reality. The simulation output corresponds to a distribution of the output variable that may be analyzed just in terms of confidence intervals. Since the actual development of the project represents a single story, a performance value located in the "tails" of the distribution may in fact occur. This possibility is enhanced in case of "fat tailed" distributions, where values far away from the central value maintain a significant probability.

The first use of simulation is in the quantification of the contingency reserve which is based on the distribution of the residual risk, that may be described for instance by the distribution of the overall project cost. The residual risk includes both the original variability of the parameters and the residual component of the major risks, *after* the implementation of the planned response actions. It's important to emphasize that the contingency reserve should be proportional, not to the expected cost but to the cost variability, i.e. to the possible lack of accuracy in estimating the total cost. The higher the uncertainty level, the greater the contingency

Fig. 14.2 Contingency size
and probability of budget
overrun

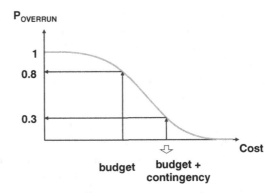

level. Figure 14.2 shows that, given the distribution of the overall project cost, it is possible to establish a relation between the total budget, i.e. the sum of the project base cost and the contingency reserve, and the probability of a budget overrun. Assuming a normal distribution for the total cost of the project, planning a total budget equal to the "expected cost" means accepting a 50 % probability of budget overrun. The expected cost represents the "zero contingency" budget, characterized by the same probability of both an overrun and under-run. Adding a contingency reserve to the expected cost reduces the probability of project overrun.

A second possible application of the simulation process is to evaluate the overall project risk, *before* planning and implementing the response actions which are aimed at reducing major risks. The project cash flow may be used as the parameter for estimating both project return and project risk (Caron et. al. 2007). Assuming the Net Present Value (NPV) as a synthetic indicator of the project financial performance, the expected NPV corresponds to the expected project return while the NPV variability, expressed by the NPV standard deviation, corresponds to the overall risk of the project. In fact, a small or large deviation from the expected NPV may derive either from single risk events (e.g. a critical test failure) or the joint impact of variations in some project parameters (e.g. a simultaneous cost increase in different commodity markets). A common measure of risk used in finance is the so called "Value at Risk" (VaR), a criteria used to check the level of risk associated to an investment. VaR was introduced to respond to questions such as: how much can we expect to lose in a day, month, or year and with what probability? Once obtained from the simulation process the distribution of the NPV, the VaR for NPV at level α, e.g. with α equal to 95 %, is the value of the NPV corresponding to the minimum result with probability α (or the maximum result with probability $1 - \alpha$).

Once the output variable, i.e. the project NPV, has been identified the structure of the simulation model may be determined. The basic elements of the model are the single cash in/cash out flows. These elements are exposed both to risk events and lack of estimating accuracy. Every deviation of the cash flows from the planned values, both in terms of size or time, creates an impact on the project NPV.

The link between the operational and the financial sides of the project is represented by the contractual milestones. The payments schedule, i.e. the release of funds from client to contractor and from contractor to subcontractors, is linked to the achievement of project milestones or progress and to the contractual terms of payment. As a consequence, during the planning stage, a forecast of the project cash flow should be obtained based on the project baseline. For instance the time and the success of a critical test may determine, through the contractual terms of payment, the time and amount of the corresponding cash inflows. On the other hand, the effort spent to achieve the milestone and the corresponding resources committed determine the time and amount of the related cash outflows. An organization that tries to avoid the risk of delays may plan its projects according to an early start schedule, which may in turn lead to relatively high expenditures in the early phase of the project and possible cash flow problems. In summary, a variation of the project NPV may derive from a change in the timing of the payment milestones, the size of the cash flows or the contractual terms of payment. In each simulation run, major risks may occur (together with their related impacts), variations of single cash flows may be sampled from the corresponding distributions and the joint effect of the different uncertainty sources on the project NPV may be calculated. For each simulation run a value of the project NPV is obtained as a function of the simulated project story. At the end of the simulation process the distribution of the project NPV is obtained and the corresponding expected value (indicator of project performance) and VaR (indicator of the project overall risk) may be calculated.

References

J.R. Meredith, S.J. Mantel, *Project Management: A Managerial Approach* (Wiley, New York, 2011)

F. Caron, M. Fumagalli, A. Rigamonti, Engineering and contracting projects: a value at risk based approach to portfolio balancing. Int. J. Proj. Manage. **25**, 569–578 (2007)

Chapter 15
Conclusions

Large Engineering Projects (LEPs) are typical examples of complex projects. Such projects need large capital investment, have long time horizons and often use non-standard technology. Moreover, they are characterized by a large number of stakeholders involved in the decision making process. For instance, for those developed in the oil and gas industry, project effectiveness is a complex measure, entailing economic performance, technical functionality, social acceptability, environmental sustainability, political legitimacy and economic development.

These characteristics require that projects find a balance between stability of the project plan and enough flexibility to adapt to emerging conditions. As a consequence, the management of LEPs has to deal with "certain" elements (issues and benefits deriving from the project's weaknesses and strengths respectively), "uncertain" elements (threats and opportunities deriving from uncertainty sources) and "unpredictable" elements. The latter may derive from the complex interactions between elements of the project itself or its environment. In fact, Project Management deals with a continuum of "certainty—uncertainty—unpredictability" or, in other words "issues—risks—unforeseen". It should be noted that the area covered by traditional project risk management, i.e. related to anticipated uncertain events, represents just a "grey area" interposed between a white area "certainty" and a black area "unpredictability".

Traditionally, the three areas have been addressed individually despite the strong dependencies existing between them by means of different knowledge areas: Project Management, Project Risk Management and Project Flexibility.

An integrated approach, which aims to cope with the overall continuum of "issue-risk-unforeseen", may use the following strategic levers:

- Apply Project Management processes
- Improve forecasting capability
- Enhance project robustness/flexibility
- Introduce real options
- Allocate, share, transfer risk
- Diversify, pool, escalate risk

F. Caron, *Managing the Continuum: Certainty, Uncertainty, Unpredictability in Large Engineering Projects*, PoliMI SpringerBriefs, DOI: 10.1007/978-88-470-5244-4_15,
© The Author(s) 2013

- Mitigate major risk
- Accept residual risk (with contingency)
- Influence project's stakeholders
- Develop a responsive organizational model.

For each of these levers a specific contribution to managing projects in condition of uncertainty and unpredictability may be identified.

The traditional approach to Project Management focuses on the stability of the project plan as a critical success factor. However, besides addressing through its basic processes the typical issues of the project such as time and cost, Project Management develops proactive measures able to cope, to some extent, with uncertainty and complexity affecting the project development. In particular the basic project control process allows for the anticipation of future issues and the proactive intervention on the work remaining in order to align the project with its objectives.

Uncertainty can be viewed as the gap between the knowledge ideally required to successfully deal with a project and the knowledge actually available. As a consequence, project predictability can be improved by exploiting all of the available knowledge. In general, the potential knowledge available to the project team may be classified in two ways: explicit/tacit and internal/external. Explicit external knowledge corresponds to data records about projects that were completed in the past, explicit internal knowledge corresponds to data records concerning the work completed on the current project, tacit external knowledge concerns the identification of similarities between the current project and some past projects in order to allow for the transferability of past data to the current project and lastly tacit internal knowledge is about possible events/trends affecting the work remaining.

In particular, the contribution given by tacit knowledge i.e. by experts' subjective opinions about the future development of the project, may concern:

- the impact from drivers that emerged during the past and consequently may influence also the future development of the current project;
- possible behaviors of the stakeholders involved in the project;
- anticipated certain/uncertain events or conditions affecting project performance in the future which may originate both internally and externally to the project;
- weak signals indicating emerging situations possibly affecting project performance.

A formal and rigorous approach aimed at integrating the contribution of the different knowledge sources is needed, in particular data records and experts' judgments. For this purpose, the Bayesian Statistics may provide helpful results. The Bayes Theorem represents a basic approach allowing an update of a "prior" distribution, which expresses the experts' preliminary opinion, utilizing the data records gathered in the field or deriving from past similar projects, in order to obtain a "posterior" distribution integrating the different knowledge contributions.

Also Project Risk Management may be considered as a way of increasing the available knowledge when dealing with a complex project. Project Risk

Management aims to safeguard the project uncertainty by moving from a point estimate of the project parameters, typical of traditional "deterministic" Project Management, toward a distribution (or range) estimate. In a sense, Project Risk Management allows us to exploit the knowledge about uncertainty which is normally expressed in probabilistic terms. In fact, each point estimate of a project parameter should be considered conditional on some simplifying assumptions that allow the elimination of the intrinsic project uncertainty. Managing this uncertainty, as in Project Risk Management, allows for a more realistic project control process.

Between contractual constraints and environmental turbulence, the project should find a trade-off between project stability, i.e. a high level of project robustness, and project adaptability, i.e. a high level of project flexibility. Project robustness refers to the properties that enable the project to respond to the possible impact of uncertain events and so minimize the required changes on previous decisions, in particular on the project plan. Project flexibility refers to the properties that enable the project to reconfigure itself, introducing and exploiting degrees of freedom into the project plan and/or the project scope. Since project robustness aims to maintain the initial project plan whilst facing changing conditions, Project Risk Management provides the most significant contribution to project robustness by means of implementing response actions to anticipated risks.

In this framework, project stakeholders are a very important source of risk. LEPs cannot be defined once and for all at their outset, rather they are shaped progressively from the initial concept by the dialectical interaction of the stakeholders involved, and the project team should try to influence stakeholders' behavior in order to align it with the project objectives. Note that projects can only be successful through the contributions made by the stakeholders, and in addition, it is the stakeholders who decide whether the project is successful. Moreover, the success criteria may be implicit and changing over time. In this context, the concept of project success appears to be inherently political. The underlying mechanism driving this political process, which are aimed at establishing the legitimacy of an interpretation of what is project success, is normally based on the stakeholders' pursuit of their interests, which can lead to coalition formation or conflict. This is an enormous challenge for the project team. A proactive strategy and action plan are required to deal with the project stakeholders in order to reduce unfavourable behaviour that might adversely affect the project and to encourage active support of project objectives. For instance, lobbying may be a way for exercising influence for or against laws, regulations or trade restraints. In summary, influencing stakeholders' behaviour means, as a matter of fact, shaping the project itself and its environment.

Besides a high level of project robustness, the increasing level of complexity and uncertainty in the business context requires a high level of adaptability to unanticipated changes. The project team should be prepared for potential unexpected events and situations that may emerge during the project life cycle. Project flexibility requires some degrees of freedom to be introduced into the project plan. In general, real options such as wait, scale, switch, expand and abandon may be exploited for a Large Engineering Project. By introducing some flexibility into the project plan, real options have the advantage of being able to postpone a decision, modify a decision in progress, reduce uncertainty surrounding it through acquisition of additional knowledge, and take advantage of volatility in its value.

Beside real options, the degree of responsiveness of the project to unexpected events reflects in particular on the critical role played by the human factor in terms of adaptability to new game rules, ability to learn from experience, interpret the emerging situation and generate and implement a suitable response strategy, instead of just triggering a preplanned contingency plan as in Project Risk Management.

In order to improve responsiveness, the organizational model should allow for a high level of diversity and independence across the different organizational units; each specialized unit maintaining its culture but developing a dialectical relationship with the others. The increasing level of interaction and communication between the units allows for the generation of innovative ideas. The project manager, as project leader, undertakes the role of integrator of the various units, becoming "the bridge" between diverse "languages".

Differentiation and interaction across the project organization are key to reacting to ambiguous situations or weak signals where the process of making sense of and interpreting the project's situation and then building consent for a response strategy becomes critical. Note that only a "mindfulness" oriented culture, focused on anticipating the emerging development of the project, can grasp the weak signals, e.g. trigger events that indicate the likely occurrence of a major risk or even a possible unexpected event. On the other hand, the typical organization's emphasis on procedures and preplanned contingency actions embodies assumptions that weaken the ability to respond to the unexpected and foster new learning. The interaction between the project team and the project context determines the interpretation of project situation, the strategy to be implemented and the related consent.

Each of the above strategic levers may be thought of as exercising an influence on the overall continuum of "issue-risk-unforeseen" and not just on a single aspect of the project. For instance, introducing some redundancy in the equipment in order to mitigate the risk of a temporary work disruption due to equipment failures also contributes to preparing the project to deal with unforeseen conditions by putting in place a reserve of resources.

Other similar measures deriving from a single lever but affecting the overall continuum of "issue-risk-unpredictability", may involve multiple subcontractors, redundancy of special equipment, standard technological solutions, robust design, no absolute beginner, modular construction, 4-D model, field engineering, open purchase orders, contractual flexibility, long term supply frame agreement, hedge currency and interest rate exposure, media exploitation to prevent social opposition, etc.

At each time along the project life cycle, the project team may use a mix of the above levers depending on the status of the project and its objectives. Note that, for a given project each lever may result in an ineffective response if applied alone, moreover, some levers may even be unavailable or may be used to a limited extent depending on the constraints affecting the project. So the choice of the mix of levers in order to obtain an effective action—what levers to use and in what measure—becomes very challenging since it requires from the project team an integrated view considering the overall effect of the lever mix on the whole continuum of "certainty—uncertainty—unpredictability" and not just on single issues or single risks.

Index

F. Caron, *Managing the Continuum: Certainty, Uncertainty, Unpredictability in Large Engineering Projects*, PoliMI SpringerBriefs, DOI: 10.1007/978-88-470-5244-4,
© The Author(s) 2013